The Science of Discovery
(why do scientists so rarely make breakthroughs)

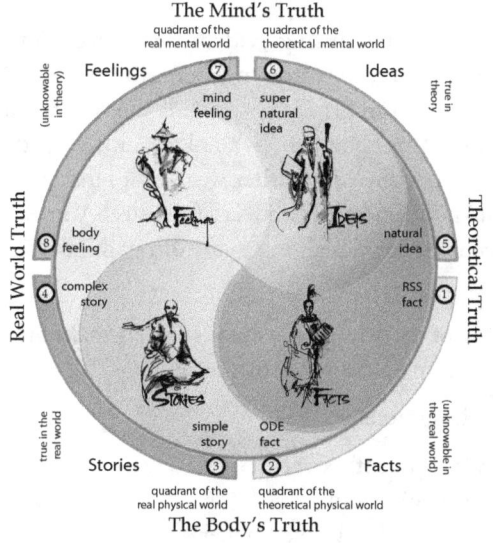

Steven Paglierani

published by
The Emergence Alliance
Nanuet, New York

© 2016, 2018 Steven Paglierani
All rights reserved.
Printed in the United States of America

This publication may not be reproduced, stored in a retrieval system, or transmitted in whole or part, in any form or by any means, electronic, mechanical, photocopying, recording, or otherwise, without the prior written permission of the publisher.

Emergence Alliance Publishing
55 Old Nyack Tpke., Ste 608
Nanuet, NY 10954

SAN 859-5380

http://StevenPaglierani.com (Musings, videos, announcements, and such. You can also download all the drawings from my books in full color.)
http://theEmergenceSite.com (This site contains the twenty plus years of discoveries which led to this book. And while I've since refined many of my terms and ideas, some may find this earlier work of interest.
https://Vimeo.com/theScienceOfDiscovery (or search Vimeo for my name)
Here you'll find recent videos explaining many of my discoveries.

Paglierani, Steven. T., 1946-
   The Science of Discovery (why do scientists so rarely make breakthroughs?)
   Book III in the Finding Personal Truth series

Includes bibliographical references.

ISBN 978-0-9844895-5-8 (soft cover)
ISBN 978-0-9844895-8-9 (ebook)
ISBN 978-0-9844895-2-7 (hard cover, jacket)

1. Science 1. Scientific Method 2. Sleep Science 2. Weight Loss Science 2. Deafness & Hearing 2. The Nature of Cancer 1. title

310 pages. 117,000 words.

| SCI043000 | SCIENCE / Research and Methodology |
| SCI075000 | SCIENCE / Philosophy & Social Aspects |

Library of Congress Control Number: 2016908373

Printed in The United States of America

Printing Number
        23 22 21 20 19 18   10 9 8 7 6 5 4 3 2

Dedication

To my father, Aldo William Paglierani, for teaching me the value in technical drawings. And to my seventh grade teacher, Dr. Clarence Branch. I hope, with this book, that I've finally become the scientist you encouraged me to become. And that my original ideas will suffice.

Thank You

To my young apprentices, Bobby and Sam, for your ever-present questions. Your youthful curiosity feeds my soul and gives me hope.

To my formal students, Carol and Karon, for your kindness, dedication, and years of gentle support. Your belief in me makes me keep reaching for more.

To my students and friends for patiently listening to my constant obsession—the science of human nature. Your years of efforts have taught me so much.

To my new friend, Reade, for reminding me to separate scientists from their method. Burdened with a flawed method, they still don't give up.

And to my friend and editor, Connie Vallone, for your years of support, encouragement, guidance, and willingness to argue with me for your truth. May you never lose your voice.

Preface to the Paperback Version of Book I: January, 2018

Recently, I began to write what I hope will become the 4th book in this series. At this point, it's been close to eight years since I published Book I. In all this time, I've not found it in me to release paperback versions. And in part, I've simply been too busy making new discoveries to find the time.

I've also had the best group of formal students I've ever had. In addition, I've begun to develop physical models which allow people to personally interact with my ideas. Mainly though, I've felt discouraged, as most people find my books too hard to read. Indeed, one man recently complained, it takes him twenty minutes to read a single page. On a good day. Even then, he said, he often feels like throwing the book at a wall.

He then added, if I'd wanted people to understand my books, I'd have made them easier to read. Logically, he's right. In truth though, I've known about this problem for some time now. Indeed, two summers ago, when I published the third book in this series, I felt certain I'd done better. But recently, one of my brightest students told me, Book III is the hardest of all.

The thing is, when I parse my books through reading difficulty software, it tells me they are written at a fifth grade level. So what's going on here?

Logical Answers Which End (or avoid) Suffering

Just prior to publishing the third book, I stumbled onto what at first seemed to be just another piece of the puzzle. It involves something I've longed to understand: how a child's developing mind decides what to pay attention to and retain. In part, this learning resembles how we come to feel alert in dark rooms and high places. But in this case, our minds create a set of filters through which all life experiences must pass.

Experiences which pass through these filters get seen as important. They alone get processed and retained. Unfortunately, much of what's in my books doesn't pass through these filters. So even when people do momentarily comprehend what I write, moments later they remember nothing.

Where do these filters come from? They likely result from a common, early childhood event—a day wherein mommy or daddy suddenly got upset then yelled, "why did you do that?" If this kind of event startles you—even once—these filters get installed. Afterwards whole categories of life experience become hard to process or retain.

What makes this happen? These parents are demanding *logical* explanations. Children cannot understand logic until they can tell time. Most of us learn to tell time roughly at around age seven. So when a parent threatens to punish a three-year-old if they cannot logically explain why something happened, they panic. This panic then biases their minds. From then on, they know, "To avoid punishment, I must always have answers." The thing is, they're not old enough to use logic to find answers. So this makes them start doing the next best thing—they learn to fabricate answers. And herein lies the problem with reading my books.

Books filled with "*logical answers* to avoid *suffering*" become best sellers—even if most of what's in these books is totally fabricated nonsense. But books which focus mainly on the nature of things—and not on logical answers to suffering—are hard to read. Even if what's in them is true.

Why *suffering?* Parents do not suddenly scream, "why did you do that?" when good things happen.

Why *answers?* "Why did you do that?" is a demand for an answer.

Why *logic?* There are just two kinds of "why" questions—*natural* and *logical*.

Natural why questions request *descriptions* of natural processes. And the form these answers most often take is, "this, then this, *then* this."

Logical why questions request *logical explanations* for natural processes. And the form these answers most often take is, "this, then this, *therefore* this."

No coincidence, the three authorities we turn to most voice their work in the second manner. Here, science focuses on answering questions about how things do and don't work. Medicine focuses on finding and fixing what's wrong; on ending suffering. And psychology focuses on logically explaining who we are and why we suffer. And in truth, this should not be a surprise.

Scientists, doctors, and psychologists were once children too.

So again, why are my books hard to read? Because the human mind is biased to pay attention to—and retain—only *logical answers which end (or avoid) suffering*. My books don't focus on suffering. They describe the nature of us and our world. This means your mind will likely filter out most of the content in this book. And the little that does make it through will get quickly forgotten.

So Can You Retain What You Read Here?

So here's the big question—can you retain what you read in this book? If you slow down, then focus on three things, you can improve your chances.

One, you'll need to focus on discovering *new lines of questioning*, rather than on finding answers. Two, you'll need to focus on finding *what's right about this book,* rather than what's wrong. And three (and hardest of all), you'll need to focus on *learning to observe the nature of things*, rather than on finding logical explanations for things. Especially this book's main ideas: logical geometry, tipping-point based math, and the new method.

Is this a lot to ask from a reader? Admittedly, it is. But as you've no doubt found out, finding personal truth is never easy. Don't give up.

Table of Contents

Opening Thoughts
The Science of Discovery
(Reader's Guide to Book III)   529

- Foolish Me. I Thought It Would All Fit in One Book.   529
- Writing Books About Discoveries is Hard   530
- Science and Discovery as this Book's Key Terms   532
- Finally, Before We Begin . . .   532

Section 1 - Introduction
Section 1 - Discovery
(what can we learn about . . . ?)   533

- Science's Holy Grail: Discoveries   533
- Science, Hopelessness, and Inspiration   534
- Answers or Questions: Which Should Science Seek?   535
- Can the Socratic Method Close Minds?   536
- Science's Big Problem: Understanding Change   537
- What Do You Get if You Marry Geometry & Logic ?   538

Section 1 - Chapter 7
Logical Geometry

(discoveries have a shape)	539
Logic, Geometry, and a New Scientific Method	539
What Makes Someone a Scientific Genius?	540
Can You Find the Tipping-Points in These Questions?	542
Can You See How the Map Emerges?	543
Can Science Make Its Definitions Clear and Precise?	544
A Few Tangible Examples of Better Scientific Definitions	545
Can Definitions Be Both Universal & Mutually Exclusive?	546
Can We Eliminate Scientific Bias?	546
Can We Get Rid of Soft Sciences?	546
Should Science Even Try to Define Spiritual Concepts?	547
Why Use Complementary Opposites to Define Things?	548
Can We Learn Anything Else From the Map of Being & Doing?	549
So How Does Using Logical Geometry Benefit Us?	550

Section 1 - Chapter 8
How the Six Logical Geometries Emerge (a brief look) — 553

Single Points (the First Logical Geometry)	554
"Connections Between Single Points" IS What You Discover	556
How Serious Can It Be to Miss a Single Point?	557
Can Single Points Really Cause Personal Problems?	558
Can Single Points Make People Blame?	559
Success, Failure, and Connecting Single Points	560
If Single Points Lead to Such Problems, Why Use Them?	561
Linear Continuums (the Second Logical Geometry)	562
Fractal Continuums (the Third Logical Geometry)	564
Can Science's "Real World Measurement" Problem Be Solved?	565
Can the Third Geometry Open a Closed Mind?	566
Can Science Actually Measure the Openness of a Mind?	567
How Do I Define An Open Mind?	568
Can Fractals & Nonlinearity Be the Key to it All?	569
What Are *Fractals* Anyway?	571
Are All Discoveries, *Fractal*?	572
The Third Geometry: the Nerd's Description	573
Can Fractal Continuums Open Minds?: Part 2	573

The Fourth Logical Geometry: Tritinuums	575
A Rationalist Scientist Gathering Data	576
A Rationalist Scientist Processing Data	577
A Rationalist Scientist Solving a Puzzle	578
A Big Question: Where Do Sensations Come From?	579
Are All Five Senses Based on Tritinuums?	581
Do Tritinuums Actually Connect the Mind to the Body?	582
Can Tritinuums Be the Source of Perception Itself?	583
Can Tritinuums Identify—and Heal—Wounds?	584
Can Tritinuums Firm Up the Soft Sciences?	587
Can Tritinuums Eliminate Scientific Bias?	588
Can "Good Things," Close Minds?	589
Can Tritinuums Reveal the Good in Negative Things?	590
Using the Fourth Geometry to Clarify Goals	591
The Fifth Logical Geometry: Crossed Continuums	592
Constellating These Observations into Scientific Patterns	594

Section 1 - Chapter 9
Using Logical Geometry
(mapping the natural world)   597

What are the Four Map-Making, Action Steps?	597
Create the Fifth Geometry: the First Action Step	597
Create the Sixth Geometry: Action Steps Two, Three, and Four	597
Mapping the Four Directions: Action Step One	598
Mapping the Four Directions: Action Step Two	598
Mapping the Four Directions: Action Steps Three & Four	599
Must All Maps Physically Parallel Each Other?	601
Can the Four Arithmetic Operations Be Mapped?	602
Mapping Arithmetic: Action Step One	605
Mapping Arithmetic: Action Step Two	605
What are Psychophysical Parallels?	606
What Would a Map of All These Steps Look Like?	607
What Has the Map Taught Us So Far?	608
Mapping Arithmetic: Action Step Four	609
Mapping Arithmetic: Concrete vs Abstract States	612
Career Advancement: What's the Best Path?	612
The Career Advancement Map: the Back Story	613

Career Advancement: How Do We Test the Guide Map?	615
Career Advancement: What's the Best Next Step?	615
• U / Bs—Uneducated / Bad Fit	615
• U / Gs—Uneducated / Good Fit	616
• E / Bs—Educated / Bad Fit	617
• E / Gs—Educated / Good Fit	618
Career Advancement: What's the Real World Path?	618
What If You Can't Find a Map's Two Questions?	619
So How Could We Benefit From Logical Geometry?	620
The Old & New Sciences: What's Different?	620
The Current Scientific Method	621
The Constellated Science Method	621
Difference 1: How You Decide Which Evidence to Consider	622
Difference 2: How You Gather This Evidence	622
Difference 3: How You Arrange The Evidence You've Gathered	623
Difference 4: How You Arrive At Your Conclusions	623
A Few Closing Thoughts About Two Methods	624

Section One - End Notes (Ch. 7 thru 9)
Afterthoughts & Resources

Notes Written in the Margins of Chapters 7, 8, & 9	627
How This Book Parallels Descartes' Discourse on Method	627
Cartesian Coordinates & Quadtinuums: a Few Last Words	628
Is Shelf Life (unchange) a Test for Toxicity?	628
Do Continuums Create Techniques or Tools?	629
On "Being" Right, and "Doing" Wrong	629
On These Maps as an Artificial Intelligence	629
On Previous Theories of Scientific Reality	630
Logical Geometry as a Formal Mathematics	630
IMPORTANT: Must You Parallel All Points to Make a Map?	631
On Fractals as Nature's Linearity	632
Resources for Section 1—Logical Geometry	**632**
On Geometry as a Path to Truth	632
On Shelf Life as a Real World Test for Quality	633
On the Efficacy of St. John's Wort	633
A Study Which Claims St. John's Wort is Ineffective	633
A Study Which Infers St. John's Wort is Effective (given you combine it	

with light therapy) 635
    On the Brain as a "Connectome" 636

Section 2 - Introduction
The Cartesian Process
(what can we learn about . . . ?) 637
    Hypotheses and the Cartesian Process 637
    Why Sleep, Weight Loss, Deafness, and Cancer? 638
    Why So Many Groups of Four? 638
    Constellated Science's "Four Processes" 639
    Process One—Step One: the Intuitive Step—Slate Clearing 640
    Process One—Step Two: the Material Step—Fact Gathering 640
    Process One—Step Three: the Empirical Step—Experimenting 640
    Process One—Step Four: the Rational Step—Pattern Seeking 640

Section 2 - Chapter 10
Sleep
(what can we learn about . . . ?) 641
    What Can Constellated Science Tell Us About Sleep? 641
    My Map of the Cartesian Process 644
    Cartesian Process ~ Step One: Slate Clearing 644
    Cartesian Process ~ Step Two: Fact Gathering 645
    Serotonin, Melatonin, & Neurotransmitters 646
    How SSRIs Treat Depression: an Analogy 647
    Melatonin: a Brief Look 648
    Sleep, Light, Sunlight, and Darkness 649
    Daylight is Blue Light 649
    Sleep & Air Quality: CO_2, Oxygen, Temperature, Breeze 650
    O_2 & CO_2 (oxygen & carbon dioxide): Measuring Air 651
    What About Sleep and Breathing? 651
    REM Sleep, nonREM Sleep, and Rest 652
    Is This a Better Way to Order Hypnogram Levels? 654
    How About Sleep & Dreams? 655
    A Look at Babies' Sleep States 655
    Cartesian Process ~ Step Three: Experimenting 656
    The Serotonin / Melatonin Sunglass Experiment 657

The Basics of Color Theory	657
Cheap Sunglasses	658
The Problem of Confounding Influences	659
A Solution for Confounding Influences	660
Finally, the First Sleep Experiments	661
My Experiments with Light & Waking Up at Night	662
The Hall Light / Bedroom Light Experiments	664
The Childhood Sleep-Environment Surveys	665
Cartesian Process ~ Step Four: Pattern Seeking	667
Re-Listing My General Categories	667
Cartesian Process ~ Our Working List of Opposites	668
Cartesian Process ~ Outcome: Sleep Questions	669
So What Have We Learned About Sleep … ?	672
Serotonin, Melatonin, and Neurotransmitters	672
Sleep, Light, Sunlight, and Darkness	674
Sleep and Air Quality: CO_2, Oxygen, Temperature	675
REM Sleep, non-REM Sleep, and the Nature of Rest	675
What Are Dreams and Why Do We Need Them?	676
Where Are We Going Next?	677

Section 2 - Chapter 11
Weight Loss
(what can we learn about . . . ?)   679

Why Can't We Lose Weight and Keep It Off?	679
Cartesian Process ~ Step One: Slate Clearing	680
Cartesian Process ~ Step Two: Fact Gathering	682
Are Dieting & Exercise Making Us Fatter?	682
Should We Finish Our Food or Not?	683
Why Are Young Children Naturally Thin?	683
Is Scale Weight a Predictor of Death?	684
Are Calories Real?	686
Is Metabolism Real?	687
Why Does Diabetes Change People's Weight?	687
Threads of Similarity: Type 1 & Type 2 Diabetes	688
Threads of Dissimilarity: Type 1 Diabetes	689
Threads of Dissimilarity: Type 2 Diabetes	689
Does the Timing of Insulin Production Match Our Eating?	690

Do Processed Foods Cause Weight Gain?	690
Does the Rate of Absorption Change Nutritional Value?	691
Are "Pre-Digested" Foods a Cause of Weight Gain?	692
What About Appetite, Hunger, and Fullness?	693
Anorexia, Bulimia, & Being Obsessed With Weight	694
What About Neurotransmitters and Weight Loss?	696
What About Will Power and Losing Weight?	697
Cartesian Process ~ Step Three: Experimenting	697
How Much Does Your Mind Affect Your Weight?	698
Can Eating a 4 Ounce Muffin Cause You to Gain 3 Pounds?	698
Weight Changes in Ranges, Not Pounds	700
Does Sleep Affect Our Weight?	702
Cartesian Process ~ Step Four: Pattern Seeking	703
Re-listing My Starting Categories	703
Cartesian Process ~ Our Working List of Opposites	703
Designing "Change Over Time" Drawings	704
The Power of These Drawings	708
The Nature of Dieting & Exercise: Why Are We Fatter?	708
Tracing the Appetite Transition Points	711
What's it Like When Our Appetite Awareness Changes?	711
Can You Have Chart Perfect Weight & Be Naturally Fat?	712
Does Dieting Make Us *More* Naturally Fat?	712
What About When Naturally Thin People Overeat?	714
Are Bulimia and Anorexia, Mind / Body Disorders?	714
Haven't I Just Broken My Rule & Offered Answers?	716
So What Have We Learned About Weight Loss… ?	716
Onward and Upward. Is This Beginning to Make Sense?	719

Section 2 - Chapter 12
Deafness
(what can we learn about . . . ?)   721

Are There Also Two Kinds of Deafness?	721
Cartesian Process ~ Step One: Slate Clearing	722
Cartesian Process ~ Step Two: Fact Gathering	725
[1] Physiological Hearing—How Does it Work?	725
What is Hearing?	727
The Ear Needs Power?	728

Where Does the Ear Get Its Power From?	729
What Was Different About James' Deafness?	730
So What Caused James' Deafness?	731
Could James' Deafness Have Been Healed?	732
Why Do So Many Parts of the Ear Come in Threes?	733
Is Signing the Same as Speaking?	733
[2] Hearing as Sensation: Is *Hearing*, Sensing Sound?	734
What Does Philosophy Have to Do With Hearing?	737
Is Noticing Sounds With No Meanings, *Hearing*?	738
How Hard Is It to Learn to Hear Greek Letter-Sounds?	740
[3] The Mind & Hearing: Is Understanding Sound, *Hearing*?	741
Must We Assign Meanings to Sounds to Hear Them?	743
Is "Sound Without Meaning" a Kind of Deafness?	743
Is There More Evidence Hearing Takes 2 Skills?	744
Have We Gathered Enough Evidence Yet?	746
Are We Ready to Define *Hearing* and *Deafness*?	747
What About the Medical Condition, "Word Deafness?"	748
What About Mental Sounds You Can't Speak?	749
What About Sounds With Multiple Meanings?	750
What About the Brain's Internal Conversions?	752
FM Radio Broadcasts as Medium Conversions	753
[4] Deafness and Hearing: What Do Deaf People Hear?	755
The Hearing to Speaking Feedback Loop	756
What Makes Learning Language Have a Season?	758
Do Baby Brains Begin to Learn Right Away?	761
The Early Years of a Baby's Brain: the Supermarket Story	762
Supermarket Design in Four Easy Steps	763
Creating Storage Spaces with Differently Shaped Holes	764
To Remodel a Supermarket, Must You Close It?	765
Containers, Content, and Storage Spaces	766
Force-Fitting Shaped Blocks into Already Existing Holes	766
Are Baby Brains Like Computers?	767
What Makes My Theories So Hard to Learn?	767
Cartesian Process ~ Step Three: Experimenting	769
What's the Worst Sickness a Human Can Get?	769
Is Reality, "Literal," or Do Our Brains Create Reality?	771
Is the Brain the Artist of Our Sense of Reality?	772

Table of Contents

How Does the Brain Paint Our Reality?	773
What Happens When Our Senses Don't Agree?	773
How Much of Our Reality is Real?	775
What Did I Learn From My New Language Experiments?	777
What's the Best Tool to Learn a New Language?	778
What Part Does Vision Play in *Hearing* Language?	780
Can You Lip-Read a Silent TV?	781
Have You Ever Experienced Tone Deafness?	782
Have You Ever Experienced Tempo Deafness?	782
Does the Sound of Massage Oil Change the Massage?	783
The Deafness of Imitating Normal	783
Cartesian Process ~ Step Four: Pattern Seeking	785
My General Categories Listed Once More	786
Cartesian Process ~ Our Working List of Opposites	786
Cartesian Process ~ Outcome: Deafness Questions	787
So What Have We Learned About Deafness … ?	788
Onward and Upward. Is This Beginning to Make Sense?	791

Section 2 - Chapter 13
Cancer
(what can we learn about … ?) — 793

Cancer and Feeling Powerless	793
Is Cancer Another of Science's Definition Problems?	794
Does Cancer Also Have a Baseline Pair?	794
Could This Be Cancer's Baseline Pair?	795
Cartesian Process ~ Step One: Slate Clearing	796
Cartesian Process ~ Step Two: Fact Gathering	798
Foggy Room Syndrome: Symptoms and Startles	799
Can the Mind Alone Provoke Physical Symptoms?	800
Can Some Breast Cancers Be Caused by Regressions?	802
Can Adult Cells Become Fetal Cells?	803
What Part Does the Origin Timing Play in Cancers?	804
What Causes Type 2 Cancers?	806
How Many Cancer-Causing Agents Can You Name?	806
Have You Heard of The Synergistic Effect?	808
Can Cancer Be An Emergent Property?	809
Are Tumors Failed Attempts to Form Organs?	810

Ionizing vs non Ionizing Radiation	811
What About Cell Phone SAR (specific absorption rate)?	812
So Do Cell Phones Cause Cancer?	812
Cancer Cell Migration vs Multiplication	813
Cartesian Process ~ Step Three: Experimenting	814
A Few Words on Designing Experiments	814
My Sister and the Chemotherapy Experiment	815
Dee Avoiding the Med Stop	816
Cartesian Process / Step Four: Pattern Seeking	817
My General Categories Listed Once More	817
Cartesian Process ~ Our Working List of Opposites	818
So What Have We Learned About Cancer … ?	819
Closing Thoughts on the Cartesian Process	822

Section Two - End Notes (10 thru 13)
Afterthoughts & Resources

Notes Written in the Margins of Chapters 10 thru 13	823
Chapter 10 - Sleep	823
On The Tao and Seeking Questions, not Answers	823
Chapter 11 - Weight	824
On Bulimia, Diagnosis, and Focusing on Personal Cause	824
On What Actually Makes a Food Good For Us	826
One Final Thought on Calories	827
Chapter 12 - Deafness	827
On "Mental Sound" vs "Inner Voice" and "Inner Speech"	827
Longitudinal Waves (e.g. sound) vs. Transverse Waves (e.g. water)	828
On Hearing as Conversions	829
On Deaf vs deaf (big "D" vs little "d")	829
Human Sounds Convey Emotions Faster Than Words	829
On Musical Ear Syndrome (MES)	830
What About Misophonia?	830
Chapter 13 - Cancer	830
On the Fear of Cancer	830
On DNA, Genes, Baseline Pairs, and Logical Geometry	831
Misc	832
On Focusing the Search, and On Discovering Baseline Pairs	832
On Babies as Constelled Scientists	833

Resources for Chapters 10 - 13	**834**
On the Science of Sleep	834
On the Science of Weight Loss	835
On the Science of Deafness	836
On the Science of Cancer Research	837
Misc	838

Section 3 - Introduction
The Mind
(what can we learn about . . . ?   839

Oops, Did I Do It Again?	839

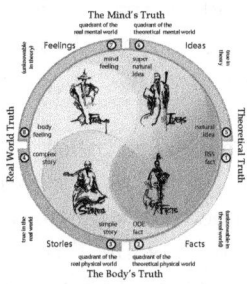

Opening Thoughts

The Science of Discovery (Reader's Guide to Book III)

Foolish Me. I Thought It Would All Fit in One Book.

Years ago, I naïvely planned to write down my discoveries. Foolishly, I thought they would all fit into one book. Somewhere during the writing of that book, I realized one book would be too large. So I broke what I was writing into what I thought would become a series of three books.

Now here I am some years later, trying to decide what to put into book three. Already, I'm overwhelmed by how much there is to say. You'd think by now I'd have realized—there is no end to what a person can discover. Until I formalized this method, I had no idea this was true—let alone why.

Then there's my discomfort with the arrogance thing. Who am I to claim I've written a new scientific method? I'm a therapist, and a writer, and a student of human nature. But a scientist? Professionally? Absolutely not.

At the same time, I keep picturing my final day in seventh grade. I'm leaving my homeroom for the last time and my teacher has stopped me to say goodbye. Mr. Branch had been both my homeroom teacher and my science teacher that year. And to my surprise, he had tears in his eyes.

I can still hear him calling me, "Peg," my nickname back then. Actually, my nickname was "Pag," but Mr. Branch was from the South. Looking back, I can't imagine how he'd managed to get that job. It was 1957 and Mr. Branch was the only black man in an all white, conservative town.

The thing is, what he told me that day is in part what's inspiring me to write this book. Nothing grand, mind you. Just an offer to write me a reference should I decide to go on to college. What surprised me though was what he said right after that.

He said I would make a good scientist.

Realize, as I begin this book, I'm closing in on seventy. I've done a lot of things in my life—some well, some quite badly. I've never seriously considered becoming a scientist before though, this despite the fact that I've loved science all my life. And as I think about it, something in my head tells me it may be too late. Can a person with no degree in science even become a scientist? For that matter, can someone close to seventy start a new profession? Or should people limit their professional aspirations to what they've gone to school for?

To be honest, I feel nervous. I expect, I'll receive some harsh criticisms. This thought makes me question if I'm doing the right thing. But when I think of the children I may affect, my head clears. If even one child discovers something, then it will all have been worth it. If more than one does, well....

My goal then will be simple—to try to awaken in you the same love of asking scientific questions which Mr. Branch woke up in me. Moreover, I'm going to begin the same way he did with me—by telling you something.

Know I have tears in my eyes.

You would make a good scientist.

Writing Books About Discoveries is Hard

When I think back to what I was feeling when I wrote my prior two books, I clearly recall feeling lost. My discoveries are so complex. How was I supposed to find the right words? In fact, right after I decided to title Book I, *Solving the Mind Body Mystery*, I realized I had no idea where to start.

Worse yet, I kept getting sidetracked by what I now call, the "definition problem." I spent the first three quarters of that book using logical geometry to define my first four terms. This meant I didn't get to the book's main theme—the mind / body connection—until well into the book.

Talk about digressing.

The problem was, I'd been worried about—and had gotten sidetracked by—Ludwig Fleck's 1935 monograph, *The Genesis and Development of a Scientific Fact*. In it, Fleck writes about science's failure to concretely define the word, *facts*. I wanted to avoid this problem. I also wanted to offer acceptable, scientific proof for my work. But the more I researched how to do this, the more confused I became.

Can you see my dilemma yet? I was writing a book about how the mind and body connect. Modern science sees topics like this one as dubious at best. To counteract this, I wanted to offer solid, scientific evidence. Unfortunately, the more I tried to learn how to do this, the more lost I felt.

Then, when I read Fleck's book, it all began to fall apart. When I read Thomas Kuhn's *Structure of Scientific Revolutions*, things got even worse. I then focused too much on defining what makes something true, rather than focusing on the subject at hand; how the mind and body do or don't connect. Ironically, I subtitled the series, *Finding Personal Truth in the Too-Much-Information Age*.

To be honest, at the time, I had no idea it was me getting lost in the *Too-Much-Information Age*. Had I realized this, I would have titled the series, *In the Too-Much-Information Age, Science Needs a New Method*.

As for what I've been calling the "definition problem," it's simple. The definition problem is what happens to people who try to define scientific terms with logic alone. The flaw, of course, is that to do this, people must use words to define words. And to see what I mean, take a moment to look up the words *fact* and *truth*.

What you'll find is that most dictionaries define a *fact* as something that is true. They then define *truth* as something that is factual. Obviously, this circular reasoning leaves these two terms undefined. Oddly, Fleck doesn't address this problem.

What he does point out though is how science treats newly discovered "facts" as if they are unchanging truths. Then at some point, science finds something better—the "new improved model." At which point, these unchanging scientific "facts" change. This makes the idea of unchanging scientific truths more resemble lover's promises to never hurt the one they love. Here, science means well. It just can't deliver.

Strangely, other than in Fleck's book, I've seen no reference to how scientific facts change—this despite the idea that this way of defining facts renders them close to useless. Know I say "close to" because there are exceptions, the most famous being $E=MC_2$. Here science uses a tangible, algebraic relationship to define its terms. Energy is defined as the technical relative to matter moving.

My point?

Science has been trying for centuries to divorce itself from philosophical discussions. One such discussion is the flawed nature of using words to describe the world. The thing is, if scientists can't find a way to address this problem, then science will forever remain flawed. A scientific method whose terms are unclear will never be good science.

On the other hand, spending three quarters of a book defining terms seems crazy. Heck, it is crazy. If only I had known I needed to write this book first. This book solves the definition problem. How? By limiting terms to "natural symmetries," such as pairs of complementary opposites—the most famous example of which happens to be *energy* and *matter moving*.

Science and Discovery as this Book's Key Terms

As for this book—*The Science of Discovery*—already, I have two terms to define. I've also inferred the pivotal question—can *discovery* be made into a science? In this book, I intend to use a new scientific method to define both terms. I also intend to offer sure and certain proof that the art of discovery can indeed become a science.

Along the way, I'll make a case for that *science*—as it's practiced today—is the business of having fortunate accidents. Contrast this with the new method which makes science the business of *connecting single points in meaningful patterns*. What the heck am I talking about? Explaining this may take most of this book. If my claims are real though, this won't matter. At least, not to science.

Finally, Before We Begin . . .

Before we begin, I have a few housekeeping details to explain. For one thing, this book's rather odd chapter and page numbering scheme. These numbers reflect my original intention; that this series was to be one book. Thus the chapter and page numbers in book three begin where book two left off.

Then there's my motive for writing this book—why in the world would I want to do this? Yes, I love science. But do I love it enough to spend a decade inventing a new method? In truth, no, I don't. But I do love children. And this is where my true motive lies.

My hope is that a revitalized scientific method will reawaken in people the joy of making discoveries. The same joy all children are born with. The joy Mr. Branch awoke in me so many years ago. Moreover, whether you're a disgruntled scientist or a struggling seventh grader, you deserve this joy. I pray this book may inspire you to move in that direction.

Finally, were you to ask me to summarize this book, I would simply say this. If you want to make more discoveries, then learn to act more like a three year old. Ask a million questions, all of which lead to new lines of questioning. And for Pete's sake, stop focusing on answers.

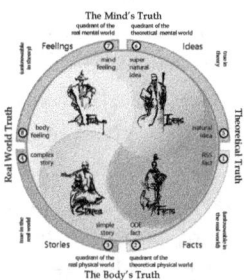

Section 1 - Introduction

Section 1 - Discovery (what can we learn about ...?)

Science's Holy Grail: Discoveries

Everything scientists do is intended to lead to discoveries. No coincidence, we reward scientists who make meaningful discoveries with things like the Nobel Prize. In a way then, discoveries are the universal holy grail of science. Yet strangely, close to 100% of the efforts scientists make fail to discover anything. Why? And why does no one ever mention this?

This is not to say scientists discover nothing, let alone nothing meaningful. Not a single life on this planet has been left untouched by science's advancements. Moreover, scientists themselves are some of the brightest, most hard working people I know. So how do we explain this close to perfect record of failures?

This book will allege these failures are due to faults in the scientific method. Some proof for this lies in the retorts people make whenever I suggest this. Indeed, most times, questioning science's methods seems tantamount to questioning religious people about God's will. To wit, when asked to discuss their method, not a single scientist has been willing to talk me through it—at least, not scientifically.

If the method is truly scientific, then why do people get so defensive when asked these questions?

Science, Hopelessness, and Inspiration

In part, this book will explore this mystery—why the current scientific method fails so often. As I've said, it's not because of anything wrong with scientists, per se. Indeed, this book will claim the problem lies entirely with the current method. Rather than bash this method however, I intend to explore a possible solution—a new scientific method—one which promises scientists discoveries *every time out*.

Is such a method even possible? Can you imagine how it would alter our lives if it was? How many more suffering people could scientists help? How many seemingly unsolvable problems might scientists solve? Moreover, can you imagine how many more children would feel inspired to become scientists? All this, just from a better method.

Speaking of *inspiration*, have you ever wondered what scientists think of this word? Not how dictionaries define it, but rather what this word means to them. Sadly, although the work of great scientists often inspires others (and has frequently inspired me), other than Einstein, few mention this word—let alone the part it plays in making discoveries.

So what is "inspiration?"

Inspiration is *experiencing the hope that you'll make a discovery*. It's the hope you'll shout "eureka," or have an aha, or discover a new lens through which to see life. Sadly, the current method fails so often that over time, most scientists lose hope. My evidence? Ask a scientist how often he or she shows up for work feeling inspired. The ones I've talked to say, "rarely, if ever," let alone that they feel inspired, day after day. And no wonder. Would you feel inspired if you failed to succeed at your job close to 100% of the time?

No surprise then that so many scientists believe in the patently false idea that discovery is 1% inspiration and 99% perspiration. To be honest, as currently practiced, 1% is wildly optimistic. Oddly, despite enduring years of drudgery and failures, most scientists never question their method. Well, almost never. Neuroresearcher, Sebastian Sung, writes in his 2012 book, *Connectome*, that no scientific method will ever guarantee answers. Talk about hopelessness. And courage. By admitting this, Sung risked the ire of an entire community who religiously swear to defend the legitimacy of their method.

How can a method with a record of close to 100% failures generate such loyalty and fervor? I have to admit, I haven't a clue. I can say that as a boy, I felt this same loyalty—and vigorously defended science myself. I also took the words of scientific authorities as absolute truths. Indeed, at times, I even thought about becoming a scientist and might have, were it not for a few of them telling me, "it's not as exciting as you might imagine." Considering what I know today, it's clear, they saw the truth about the scientific method even then.

Answers or Questions: Which Should Science Seek?

The good news is, the solution may lie in asking one simple question. The question? What should scientists be looking to discover? Obviously, the current method looks to discover explanations for unanswered questions. The one we'll be exploring in this book does as well. But the current method claims these "explanations" must lead to answers. Whereas the new method claims, focusing on answers closes minds. And obviously, closed minds cannot make discoveries.

Now take a moment to step backwards through what I've just said.

- Closed minds can't make discoveries. (Closed minds believe they already know the answers.)
- Answers close minds. (Why look elsewhere if you already know the answers? Why reinvent the wheel?)
- The current method focuses almost entirely on finding answers (and by doing this, inadvertently closes minds).
- All science begins with *unanswered questions*.

Does this sound like a Zen koan? Can you see where the change must occur? Clearly, looking for answers plays a role in any scientific method—including in the new one. But if this simple logical sequence is correct, then a truly scientific method must never make answers its focus. So while science revolves around *unanswered questions,* answers cannot be its goal.

Know we'll delve into this assertion in depth, later, when we explore the nature of human consciousness. After all, to make a discovery is to alter one's consciousness—to permanently open a part of one's mind. Indeed, one of the better ways to know you've made a discovery is to test yourself for pleasant surprise. Actual discoveries pleasantly surprise you for the rest of your life. Momentary insights please you only for moments and afterwards, feel uninteresting and hollow.

This question—*what proves you've discovered something?*—will also be something we'll explore in depth in this book. Here, I'll advocate for the idea that education should focus more on teaching kids the skill of reinventing the wheel—and less on how best to parrot answers. No coincidence, these two statements parallel problems in the current scientific method. Both education and the current method focus on finding and parroting answers.

Focusing on answers rather than questions—can this be why so many educated people can't think outside the box? Can this be why the world is so biased against people who "reinvent the wheel?" Can this also be why letters after people's names do not guarantee competence? And more, can it really be this simple? Can focusing on answers really be this bad?

This book will claim that chasing answers is the main flaw in the current method. Why? Because answers end searches and by doing so, close minds. The alternative? Constantly ask yourself how all natural things connect. Because there are endless numbers of connections within and between all natural things, asking how things connect generates an endless stream of questions. Moreover, it's said that asking the right questions can at times open closed minds. Can it? Let's consider this for a moment.

Can the Socratic Method Close Minds?

Recently, my friend Bob and I had a discussion as to the best way for teachers to open minds. He claimed the Socratic Method opened minds—I claimed it closed them. After thinking on this for several hours, I realized we both were right. Moreover, I saw that our two viewpoints were yet another pair of complementary opposites. Complete with example stories.

His story?

My friend teaches college law classes and uses the Socratic Method to get his students to consider new ideas. In this, he claims much success. Knowing him, I believe he is right.

My story?

Plato—being Socrates' loyal student and self-sworn avenger—wrote stories wherein Socrates used his method to get his enemies to publicly humiliate themselves. This didn't open their minds but rather, made them look like idiots.

My realization?

My friend was using Socrates' method with open-minded people, and this led to them having realizations. Whereas Plato had Socrates using his method with closed-minded people, and this led to them exposing their closed mindedness.

We'll talk more about the Socratic Method later on in the book, including how it both does and does not parallel the current scientific method. Certainly, Socrates's idea—to use logic to eliminate flawed hypotheses and logically impossible outcomes—is similar to the current method. And of course, accomplishing these things has much merit. Including that Aristotle attributed to Socrates the discovery of induction as a method. As such, he claimed Socrates had invented a scientific method.

The question still stands though—does the Socratic Method open or close minds? Or does it merely affect minds based on the state they're currently in? A truly scientific method must do more than just move minds in their current direction. A truly scientific method must open closed minds.

What does open minds then? Just one thing. Curiosity. Here, curiosity is *the hope of discovering connections*. The method we'll explore in this book generates an endless stream of questions about how natural things connect. As such, it generates curiosity and so, has the power to open closed minds.

Science's Big Problem: Understanding Change

Many of the ideas which define the present method come from things Socrates and the ancient Greeks said about change. Arguably, one of the more important involves the argument between Parmenides and Heraclitus. Parmenides believed all change was illusion and that nothing in nature actually changes. Heraclitus argued that the nature of the world IS change—that nothing ever stays the same.

Who do you think was right? Modern science chose to side with Parmenides. Here, unchanging *outcomes* prove things are true, and unchanging *ideas* are how you test for unchanging outcomes. Unfortunately, applied to the real world, this belief results in nothing but confusion. Unchanging laboratory outcomes—science's gold standard for proof—rarely translate to real world, individual cases. Why? Because in the real world, no two acorns, or clouds, or roses are ever the same. Nor does any acorn, or cloud, or rose remain the same even for one moment.

In large part, this is why modern psychology has had no way to measure individual people. Talk about variation. People are among the most complex natural things in the universe. At the same time, in my previous book, I described five real world measures which together delineate what makes each person that person—their core personality. None of the usual vagueness here either. Rather, a formal system so precise it can—with 100% certainty—scientifically determine which group of a few thousand or less you belong to out of seven billion people.

Imagine the predictive power of such a system? Not to mention, the good it might do in real world situations like matching teachers to students. Of course, the question is, is predictive power like this even possible? Obviously, if the method you use reconciles these two ancient viewpoints, it is. You see, it turns out, both Parmenides and Heraclitus were right—the natural world both does and does not change constantly. The trick, of course, is to discover a method which allows you to know which view to use when.

No coincidence, this is exactly what the new method does.

What Do You Get if You Marry Geometry & Logic ?

As for where we'll begin, we'll start by exploring the math which makes the new method possible. Know the opening page of the next chapter contains a brief, but concise—formal and complete—description of this math. There we'll explore what it is, what it does, and why it's the best path to scientific discovery. But please know, should maths like geometry intimidate you, relax. They'll be no formulas to memorize, nor boring tables, nor inscrutable symbols. Rather, we'll focus entirely on how a simple sequence of six geometric shapes reveals the hidden connections inherent in all natural things—including in things like IQ, thoughts, and feelings.

What is this new math like? In a way, it functions a bit like crossword puzzles. What I mean is, with crossword puzzles, the arrangement and intersections of blank spaces give you clues as to what data to use—and where to place it. The new math—logical geometry—functions similarly. In each shape, there are logical arrangements of intersection points and spaces. These points and spaces then guide your data collection, experimentation, and final placement.

What do you gain from discovering connections in natural things? In a word, you discover scientific meaning. Why? Because it turns out that all things—including all natural things—get their meanings from how they connect. What kinds of connections are we talking about? The sort that turn a certain group of seven stars into the Big Dipper. Moreover, unlike the kind of formal logic you learn in college, the logic in logical geometry is easy to learn. Indeed, babies even use it, albeit, without ever knowing.

It's what makes young children so permanently curious.

In this way, logical geometry holds the power to restore your love of learning. It also can potentially raise your IQ. Moreover, the more you use it, the higher the chances your IQ will rise. And yes, clearly, I have a lot to explain. Given what I've been telling you is true though, then the real question should be this.

What will you do with all this new energy—and how will you choose to make the world a better place?

After reading section one, you'll be well on your way.

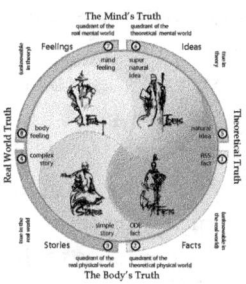

Section 1 - Chapter 7

Logical Geometry
(discoveries have a shape)

Logic, Geometry, and a New Scientific Method

In this chapter, I'm going to introduce you to the inner workings of a new scientific method. This method uses logic to arrange data into geometric patterns. Here it's easy to see why I'd want to base a scientific method on logically-arranged data. But why use geometry?

The goal of a scientific method is to discover how natural things connect. It's the patterns in these connections that give things meaning. Geometry is patterns of interconnected points. Thus arranging data geometrically allows you to discover otherwise hidden meanings.

The scientific method I'm introducing here utilizes a progression of six geometric shapes. This progression turns out to be a clear and certain path to discoveries, both scientific and otherwise. Here the points in these shapes function like place-markers for the data, and a set of logical rules tells you how to place data on points or in spaces.

In a way, this "logical geometry" augments the brain's innate ability to connect ideas. No coincidence this progression of shapes also describes the six patterns underlying IQ. It also reveals what makes someone a scientific genius. Moreover, because logical geometry charts a clear path to discovery, for the first time, scientists can make discoveries *every time out*.

Realize, once you know how to use these six shapes, there's literally no limit to what you can explore. A topic which has never made sense to you? No problem. With logical geometry you can either find its essence—or figure out where you get lost.

For example, what if you're a medical researcher exploring the progression of a strange new illness? With logical geometry, you can trace this progression backwards in time and constellate the symptoms. Or say you're a scientist investigating a tragic industrial accident. With logical geometry, you can identify the interactions critical to your search.

You can also use logical geometry to free your mind of personal problems. It's hard to make discoveries when your mind is overwhelmed. For instance, say there's a fear that's been haunting you—the fear of flying, or pubic speaking perhaps. With logical geometry, you can find—and heal—the pattern causing this fear.

You can even use logical geometry to help with parenting issues. For instance, say you've been arguing with your thirteen year old about when she can start dating. With logical geometry, you can map the argument and see where the two of you get stuck. Imagine resolving that argument.

Of course, before you can do any of these things, I'll first need to show you how logical geometry works. This is what this chapter is about. We'll begin by taking a look at the end game—the sixth geometry—the geometry of discovery. Here, we'll look at the logic which guides the data entry process. We'll then explore the progression of five geometries which together, lead to the sixth—including two rather strange arrangements called "tritinuums" and "quadtinuums."

Overall the thing to keep in mind is that I have one goal here—to enable you to make scientifically sound discoveries *every time out*. And yes. I know. This is a bold claim indeed. But if, while I'm presenting my evidence, you can embrace the spirit of science and keep an open mind, you'll soon be on your way to making your own discoveries.

What Makes Someone a Scientific Genius?

If science's ultimate goal is to discover the patterns in natural things, then you'll need a proper method to help you to see how things connect. In "constellated science"—the scientific method I'm introducing in this book—it's the sixth geometry which allows you to see these connections.

In effect, this geometry functions like a scientific template. It tells you where to put things and when things are still missing. Each time you place something into a "map" then, the magic happens. Once placed, this geometric arrangement reveals previously hidden relationships in this data.

Logical Geometry (discoveries have a shape)

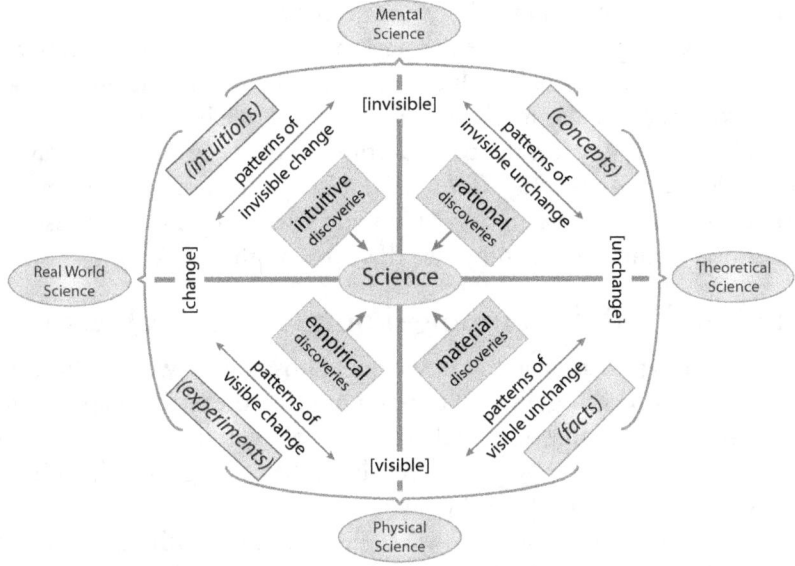

Is this thing visible? (vertical axis question)
Does this thing change? (horizontal axis question)

A similar ability exists in the minds of scientific geniuses. What hasn't existed previously is the knowledge of which patterns underlie this genius. In the diagram I've placed above, I've used the sixth geometry to map these patterns. And in part, these patterns include the way the geometry defines terms—"real world science" as the complementary opposite of "theoretical science," for instance.

More important though is the way in which this geometry reveals how all these terms interconnect—as well as revealing what's the same and what's different, all the while leaving nothing out. So what do I mean by "genius?" Ostensibly, this is a hard word to define. A lot of very smart people have struggled to define it and failed.

According to this map though, *genius* is "the ability to find the patterns in already existing things." Interestingly enough, this same map also reveals how genius and creativity relate. Here, *creativity* is defined as "the ability to destroy things in order to discover new patterns." It seems then that *creativity* is the complementary opposite to *genius*. Who knew?

Were you just surprised by how elegantly this map defined these two terms? Were it not for logical geometry, I'd never have discovered this beauty. Moreover, because logical geometry precisely parallels the patterns present in all natural things, it gives you an unlimited ability to define things like the nature of creativity and scientific genius.

Now let's begin to explore the design of this map. Start by noticing the cross in the center of the drawing. Like all crosses, this one is made up of two perpendicular lines—one vertical and one horizontal. Nothing unusual here. Now notice the two questions at the bottom of the drawing. The vertical axis question—*Is this thing visible?* And the horizontal axis question—*Does this thing change?* Can you guess why I chose these two questions. A hint. It's much harder than you might think.

Can You Find the Tipping-Points in These Questions?

In truth, expecting you to know why I chose these two questions is a bit unfair. For one thing, I've yet to point out that the entire map emerges from just these two questions. I also haven't told you this map emerges in an unbroken sequence of logical steps. From beginning to end, each part of this map logically leads to the next.

Know this sequence of unbroken logic guarantees nothing in this map can be ad hoc. Every single element in this map is logically related to every other element. Of course, the real question is, how can such a complex shape emerge from just two questions? The answer? These questions aren't normal questions. They're both tipping-point based. What I mean is, each question has only two possible answers—each the complement of the other. Here the Vertical Axis has two answers: *visible* or *invisible*. And the Horizontal Axis has two as well: *change* or *unchange*.

Obviously, both of these questions can tip in only two directions. This means they cannot result in ambiguous answers. Moreover, this is true even when making real world measurements. Thus you can always rely on the outcomes of these two questions—either something is visible or invisible, and it's either changing or not changing.

Now ask yourself how often the present scientific method can live up to these three standards. [1] Completely clear questions. [2] Completely measurable answers. And [3] outcomes which are reliable 100% of the time. In truth, nothing the present scientific method does can come close to passing these tests. Know we'll talk at length later about why.

First, let's have a look at how such a complex geometry can emerge from just two questions. We'll begin with a five-cent tour of the order in which these elements emerge.

Can You See How the Map Emerges?

Imagine I've asked you to answer this map's two **questions**. Then imagine I've placed the four possible **answers** at the ends of their respective axes. Can you find where I placed these **answers**—*visible* and *invisible*, *change* and *unchange*—in the map? Know these four patterns are what all natural things have in common. No surprise it took me years to discover them, along with the two questions which place them into the map.

Now imagine I combine **adjacent answers.** Then imagine I place each of these composite answers into a quadrant. These quadrants comprise the four kinds of discoveries scientists can make—*material discoveries* (patterns of visible-unchange), *empirical discoveries* (patterns of visible-change), *rational discoveries* (patterns of invisible-unchange), *and spiritual discoveries* (patterns of invisible-change). They also refer to the four kinds of things discoveries can be based on (e.g. data, theories, hunches)—as well as the four methods scientists can use (e.g. empirical experiments, rational designs).

Know that in previous books, I referred to these four classes of discoveries as the wise men's four truths. There I called them *facts, stories, ideas,* and *feelings.* Here, feelings were defined as *invisible change*, stories as *visible change,* ideas as *invisible unchange,* and facts as *visible unchange.* But regardless of what you call them, they still refer to the same four classes of patterns. And yes, "unchange" is an odd word to use. But I haven't found a better word—and we call things which don't change, "unchanging."

Finally, imagine I now combine **adjacent classes of discoveries.** Then imagine I place labels for these **four natural sciences** on the four outer edges of the map. Here, the two kinds of discoveries on the left side of the map—the two which describe things that change (stories + feelings)—define the natural science we refer to as *real world* science. Whereas the two kinds of discoveries on the right side of the map—the two which describe things that do not change (ideas + facts)—define the natural science we refer to as *theoretical* science. And the two kinds of discoveries at the bottom of the map—the two which describe things we can see (facts + stories)—define the natural science we refer to as *physical* science. Whereas the two kinds of discoveries at the top of the map—the two which describe things we cannot see (ideas + feelings)—define the natural science we refer to as *mental* science.

Are you beginning to see why I call this map, the map of scientific genius? In part, because this map defines everything a scientist could possibly explore. For example, say you wanted to explore something complicated, like Heisenberg's *Uncertainty Principle.* If the map of scientific genius is real, then the two questions which define this map should parallel the two patterns Heisenberg uses in his formula. And they do. *Visibility* equates to stillness *(position / particle / time)*. And *change* equates to movement *(momentum / wave / energy)*.

In effect, all things in the physical world share only two fundamental patterns. *Visibility* (Is this thing visible?)—and *change* (Does this thing change?) To me, this kind of elegance never ceases to amaze me. With just a few steps, you arrive at a template for all possible scientific discoveries.

Admittedly, you'll need to learn to set this unfolding process in motion. You'll also need to learn to use the map's logic to insert data. Once you do though, you'll be well on your way to making your own discoveries. Hopefully, the following examples will get you started.

Can Science Make Its Definitions Clear and Precise?

One thing the present method fails to offer is a way to define its terms. For instance, consider the word *facts*. Incredibly, it offers no definition for this word. Contrast this with constellated science's definition which is measurable—even in the real world. Constellated science defines *facts* as *visible unchange*. (For a comprehensive explanation which offers proof for this, see Book I in this series.)

How do you arrive at this kind of precise definition? You plug your data into the prototype map. So say you wanted to define one of science's more nebulous terms—the cosmologist's phrase, *black holes*. With constellated science, you'd start with the prototype map's two questions, then use these questions to plug data into the prototype map.

So are black holes *visible*? Obviously, not. We know they exist only because they effect what we can see; the stuff around them. And do black holes *change*? Obviously they must—the effect they have on other things constantly changes. Together, these two answers then tip black holes into the upper left quadrant. They're invisible—thus they exist in the natural state we call mental science. And they change—thus they exist in the natural state we call real world science. Moreover, despite how odd it might sound to call black holes a real world / mental science (*invisible change*), in truth, this reference is just an elegant way to define something only the mind can "see"; *energy*.

What about the idea that the content of this quadrant is also referred to as *feelings*? How can black holes have anything at all to do with feelings? To begin with, the map defines *feelings* as *invisible change*. Thus feelings can be mental and/or physical. And were you unfortunate enough to be near a black hole, this is literally the essence of what you'd experience.

Now we'd need to switch our focus from the experience to the concept. So while "the concept of black holes" also refers to something

Logical Geometry (discoveries have a shape) 545

invisible, all concepts are defined as *invisible unchange*—they're unchanging generalizations. Thus even if we never encounter a black hole, in theory, it's always possible they exist. And once a concept exists, as a concept, it never changes. (Again, for a thorough discussion of why, see Book I in this series).

For example, take the idea that "lamps light rooms." Even if no lamps or rooms still existed, the idea that "lamps light rooms" would remain. The same goes for the idea that "pencils write" and "rain falls." As ideas, these two things will always exist and never change. This means, while the experience of black holes goes into the upper left quadrant, the concept of black holes would go into the upper right quadrant. Thus black holes are an example of theoretical / mental science; a rational truth; an *idea*.

A Few Tangible Examples of Better Scientific Definitions

Now let's see what we can learn about something more tangible, say "an already-washed load of laundry." In what quadrant would you place a finished load of laundry? To begin with, you usually don't wash laundry to improve your spiritual condition. Nor do you wash it to discover new ideas. The entire point of washing laundry is to get it done. Moreover, when it's done, you can see that it's done. Thus done-being-washed laundry is visible.

As for whether it changes, according to the dictionary, it doesn't. After all, this is what the word "done" means—it means something's finished changing. This means done-being-washed laundry is a physical thing (visible) which exists only in theory (unchanging). Thus it goes into the lower right quadrant—as theoretical / physical science—as a *fact*.

The thing is, according to the prototype map, facts exist only in theory. So how can done-being-washed laundry—or any fact, for that matter—exist *only* in theory? Before I explain this, we first need to take another look at laundry. Only this time, we need to look at it as it's being washed.

Where in the map would we place "being-washed" laundry? To find out, once again, we need to ask the two quadrant-locating questions. Question one. Can you see laundry as it's being washed? In other words, is this kind of laundry visible? Clearly, it is. And question two. Does it change? Well, we certainly hope it does. After all, this is the whole point—changing dirty laundry into clean laundry. This means being-washed laundry exists in the real world, physical state. Thus it tips into the lower left quadrant, the quadrant of empiricism and *stories*.

No surprise this description exactly fits the essential nature of being-washed laundry. It's the story of how a load of laundry gets done. As for referring to done-being-washed laundry as existing only in theory, this explains why in real life, we often feel it's never done.

Can Definitions Be Both Universal & Mutually Exclusive?

Now review the four examples we just talked about. Did you notice how each datum can fit into only one quadrant? In other words, if you correctly place content, it's not possible to place the same content into any other quadrant. At least, *not in this particular form*.

The thing is, I've also somehow managed to insert references to black holes and laundry into two quadrants each. How is this possible?

In truth, anything entered into the map can be restated so that it fits into the three other quadrants. In other words, all data can be expressed materially, empirically, rationally, and spiritually—as facts, stories, ideas, and feelings. This means, while the content of each quadrant is mutually exclusive (data which fits into one quadrant can't fit into any other quadrant), all things can be viewed from all four viewpoints. They are all also universal.

Can We Eliminate Scientific Bias?

This idea—that all things in a map can potentially be expressed in all four forms—raises an important point. By inference, in order to present a scientifically comprehensive picture, what's being explored must be complete. Moreover, to be complete, anything entered into a map must be expressed in all four forms. If not, then it's measurably biased.

Now if you're like me, in and around fifth grade, you were taught that science is unbiased. Am I saying the present scientific method is biased? Duh! Did you ever ask a scientist to explain the feeling of being conscious? Even if you set aside the present method's bias against feelings per se and it's bias toward material things, how can you explain science's claim that it's neurons all the way down? We can't measure consciousness, so it's not real ? Thus feeling conscious is just an illusion?

The point is, if you enter all four forms of the things you're exploring, your scientific conclusions cannot be biased. Nor incomplete. And lest you think this is a minor point, ask yourself this. How many things can you name in science which are truly unbiased, while leaving nothing out?

Can We Get Rid of Soft Sciences?

Speaking of leaving things out, how about the types of science most scientists pooh-pooh as being soft sciences, such as human behavior. For example, say you wanted to explore the meaning of the word *smiling*. What does this even mean?

So say you're referring to that you saw your friend Jim smile in a particular place and time. Here, smiling is visible and unchanging. Thus it's an ODE fact (*an observed definite event*) and belongs *only* in the material discoveries quadrant. But if you're referring to the time your wife smiled because you came home, walked up to her, and pulled flowers from behind your back, then smiling is visible and changing. Thus it's a Simple story (*a sequence of three or more related facts*) and can *only* go into the empirical discoveries quadrant. However, if you're generalizing about how men smile, then smiling is invisible and unchanging. Thus it's a Natural idea (*a generalization drawn from a group of related stories*) and belongs *only* in the rational quadrant. But if you're referring to what you experienced when you saw your eight year old daughter play cello at her first recital, then smiling is invisible and changing. Therefore it's a Mind feeling (*a mental observation which exceeds your capacity to define with facts, stories, and ideas*) and belongs only in the spiritual discoveries quadrant.

Now consider how using all four quadrants forced you to account for the entire nature of smiling? Taken together, these four examples truly present a comprehensive and unbiased picture of this word. The thing is, for constellated science to be considered a real science, it must be able to do this even for words which are vague and nebulous. So can it define things people do during spiritual practices? Moreover, will we discover anything if we do this? Let's find out.

Should Science Even Try to Define Spiritual Concepts?

When people are learning to meditate, they frequently get told to "be, not do." Indeed, I remember struggling to understand this obviously strange advice. Moreover, anyone who has studied meditation knows that when it comes to words, dictionaries don't help. Sadly, neither will most meditation teachers—they're too biased towards the spiritual aspects of meditation. In effect, to them, meditation falls outside of the realm of science.

So if we use the sixth geometry to map these two words—*doing* and *being*—can we scientifically define *meditation*? If constellated science is truly scientific, we can.

We'll start with the word *doing*. Where does the word *doing* go into the map? To begin with, we use the word *doing* to refer to one of two things. One. To the actual act of doing something. Two. To the idea of getting things done. And in the case of meditation, the word *doing* refers to the latter—focusing on getting things done. Accomplishing things.

This is why we call lists of what we'd like to accomplish, "to-do" lists. And why, when we finish things, we treat them as if they cease to exist—by crossing them off our lists. It's also why we refer to these crossed-off-the-list things as *done*. In effect, by "done," we mean they're "done changing."

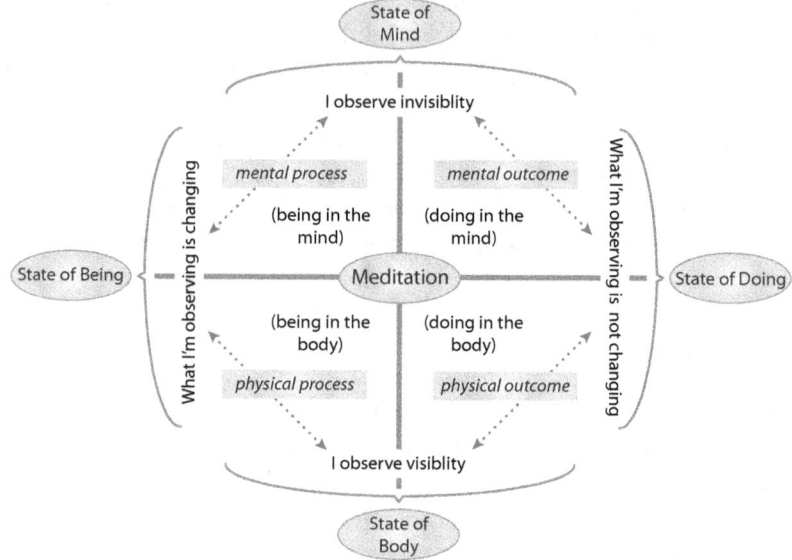

Vertical Axis Question:
Is what I'm observing visible?

Horizontal Axis Question:
Is what I'm observing changing?

The problem is, the prototype map defines real world things—atoms to atmospheres, plums to plummeting stock prices—as things which constantly change. So what do we mean when we say things are "done?" Do we mean they are finished changing? According to the map, this can't be true. Yet at the same time, the idea that nothing gets *done* sounds crazy. So have we just found a fault in the sixth geometry? Not really. And to see why not, consider what the sixth geometry does to define terms. It uses geometric arrangements of logically complementary pairs. In other words, a thing gets defined by comparing and contrasting it to its perfect opposite. So what is the complementary opposite to "doing"? And how will identifying "doing's" opposite scientifically define this term?

Why Use Complementary Opposites to Define Things?

Start with this. If nothing actually gets "done" in the real world, then what are we doing when we try to get things done? According to the map, we will always be somewhere in the process. We're always *being*, not *doing*. Now if this sounds like nonsense—or mere semantics—then ask yourself this. Why can't you reach the end of a rainbow? Do you know? For those who understand the science behind rainbows, this is obvious. Rainbows exist only in relation to where you're standing. This means, each time you move, the end of a rainbow will move with you. This makes the end of a rainbow a moving target which will always remain out of reach.

The same idea holds true each time you try to get things done. Done things exist only in relation to where you are standing—where you are now. Each time you move, your end point will move as well. This is why we always feel like there's more to do. In the real world, there always is.

Why then do we feel at times like we've accomplished a lot? In truth, this *feeling* has more to do with how conscious we are during the "being" process than with how much we accomplish. For example, say you spend the day cleaning your basement—but do this work mindlessly. In these cases, most people feel like they got nothing done. But if you take ten minutes to consciously clean a single kitchen drawer, most people feel they've accomplished a lot.

Are you still having a hard time wrapping your head around this idea? Then consider some real world examples. For instance, say your kids have gone to bed and you've finally finished picking up their toys. How long will this stay done? Or say you've finished mowing the lawn, or helping your daughter with her math homework. Will these things stay done? Or say you've given your dog a bath, or washed your car, or eaten breakfast, or paid your bills. Will any of these things ever stay done?

This explains why we sometimes work hard all day and feel we've accomplished nothing. We were so focused on the goal that we have no ability to recall the process. Moreover, that the map so clearly defines this experience points to how powerful it is. Truly scientific definitions. Imagine that.

Can We Learn Anything Else From the Map of Being & Doing?

At this point some might argue, "it's true, these things do change. But at one time, they were also done." And in theory, these people are right. In theory, they were done. But in the real world, since neither time nor place ever stands still, nothing can actually be "done." This makes *being* and *doing* two sides of the same coin. In other words, they're complementary opposites. Here, *being*—the process of witnessing change—is a state which exists *only* in the real world. Whereas *doing*—the done-changing outcome—is a state which exists *only* in theory.

Moreover, like trying to see both sides of a coin, it's not possible to simultaneously see *being* and *doing*. At any given time, we can witness only one of these states. And yes, this is not how we normally think about these two words. But if you use the horizontal axis question—does it change?—to plug *being* and *doing* into the map, it all becomes clear. *Doing* is like trying to reach the end of a rainbow. And *being* is like watching a rainbow. And all it takes to see this is to ask one quadrant-locating question—does what we're talking about change?

This then is a good example of how the sixth geometry can scientifically define even vague, nebulous words. Here *being* is the real world experience of witnessing change. And *doing* is the theoretical experience of witnessing no change. So when meditation teachers advise us to *be, not do*, all they're saying is, sit still and just "witness change."

Is this sinking in? This is it, the whole meditation enchilada.

Meditating is "witnessing change."

So much for needing twenty years in order to understand meditation.

So How Does Using Logical Geometry Benefit Us?

In a moment, we'll begin to explore the sequence of five geometries which lead up to the sixth. Before we do, let's briefly review what's been said so far, starting with the purpose of the sixth, logical geometry.

The point of the sixth geometry is to guide and define the scientific discovery process. Here the points and spaces in this complex-geometric shape act as visual and logical guides. These guides organize the data gathering and the data entry process. They also show when data is missing—and when the map-making process is complete.

More specifically, the sixth geometry works by visually and logically mapping words and data into scientifically-significant patterns. In doing so, it reveals the relationships hidden within this data. Once visible, these relationships then reveal how the subject of this map contrasts and compares to all other natural things. In doing so, it scientifically defines the place this thing occupies in the natural world.

What is the actual process like? There are only four steps. Collectively they are called, the "unfolding process." In Step One, you create and ask the two map-defining questions. In Step Two you assign the answers to these questions to the four axis points in the map. In Step Three you combine adjacent answers to define the four quadrants. And in Step Four, you combine adjacent quadrants to define the map's natural boundaries.

In the prototype map, the two map-defining questions are—*Is this thing visible?* and *Does this thing change?* Like all map-defining questions, these two questions have several things which make them special. For one thing, these questions function like coin tosses, in that they are tipping-point based. In other words, each question has but two possible answers (visible / invisible, change / unchange), and these answers are logically and geometrically, complementary opposites.

This idea—that each question's answers are *complementary opposites*—is a big part of what makes this process scientific. Literally, it means each answer contains none of the other. Yet together, the two answers comprise all possibilities. Scientifically, this guarantees several things, including that:

- **The Discovery Process Will Contain No Ambiguity**: these questions—and all terms used—will be clearly defined.
- **The Discovery Process Will Be Unbiased**: these questions—and all terms used—will always account for all possibilities.
- **The Discovery Process Will Be Pragmatically Testable**: these questions —and all terms used—will lead to practical, real world tests which yield accurate, real world answers.
- **The Discovery Process Will Employ No Statistics**: the results of all terms and tests will be 100% certain, even in the real world.

As for how you arrive at a map's two-defining questions, know we'll spend much of this book learning to do this. As I've said, it took me years to find the two questions in the prototype map—even longer to figure out where to place the four answers. Fortunately, once I'd done this, the rest of the map unfolded mostly on its own. Like finishing a puzzle, the map's logically geometric "empty spaces" guided the rest of the process.

Once done then, this map became the template for every other map I've made. For instance, I used it to know where to place things in the map I just introduced you to—the map of Being and Doing. Here, I asked myself where *being* and *doing* would best parallel the prototype map. And I quickly realized that the state of *being* parallels real world science, in that the defining quality of real world science is "change." So *being* is focusing on change. I then checked this by asking myself if *doing* paralleled real world science's complementary opposite—theoretical science. And it does. The defining quality of theoretical science is unchange—and the state of *doing* is focusing on something that is done changing.

Obviously, there is much left to learn before you'll be ready to create your own maps. The good news, though, is that the hardest part—defining a scientifically sound method—has been done for you. So once you learn to use the unfolding process—and once you learn how to find complementary opposites—every mystery will motivate you towards more discoveries.

The Creation of the Four Sciences

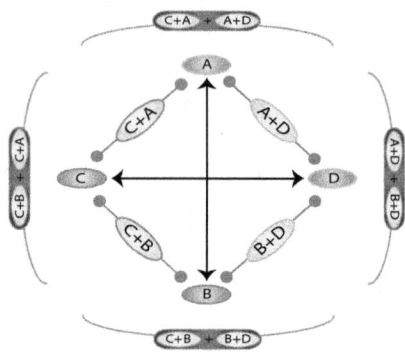

The two questions become
The Four Prototype Answers

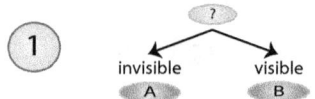

Question One (the vertical axis question)
Is this thing visible?

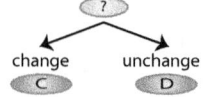

Question Two (the horizontal axis question)
Does this thing change?

①

The four pairs of adjacent answers become
The Four Kinds of Scientific Discoveries

②

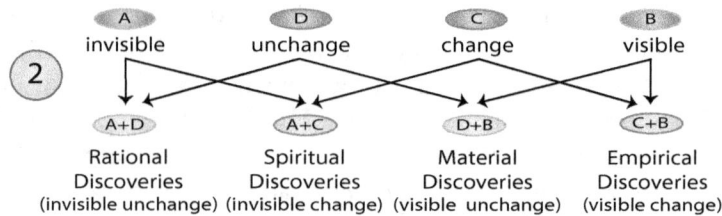

The four pairs of adjacent discoveries become
The Four Classes of Scientific Endeavor

③

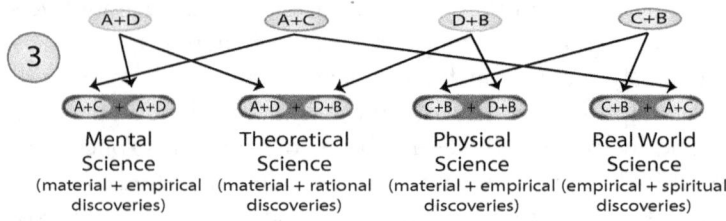

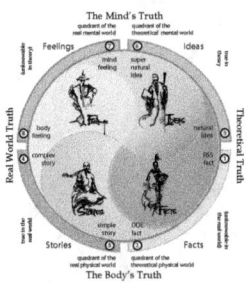

Section 1 - Chapter 8

How the Six Logical Geometries Emerge
(a brief look)

[1] **Single Points** *(the Genesis Tool)*
[2] **Linear Continuums** *(the Measuring Tool)*
[3] **Fractal Continuums** *(the Possibility Defining Tool)*
[4] **Tritinuums** *(the Consciousness Creating Tool)*
[5] **Crossed Continuums** *(the Expanding Tool)*
[6] **Quadtinuums** *(the Discovery Tool)*

Single Points
(the first geometry: the Genesis Tool)
© 2008, Steven Paglierani, The Center for Emergence

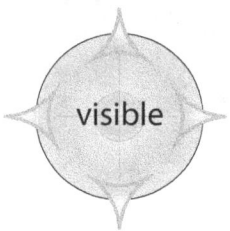

a single point,
with no direction and no connections

Single Points (the First Logical Geometry)

How can a better scientific method emerge from such a simple process? In part, it's because logical geometry forces you to use your whole mind. The prototype map contains all six logical geometries. Thus this map accounts for all possible mental patterns. In doing so, it literally has the power to reveal *all possible discoveries in all possible minds*.

Another factor concerns how the map affects IQ. IQ is the measure of the mind's ability to recognize patterns. Since the map accounts for all possible mental patterns, in theory, it could raise your IQ. Of course, for this to happen, you must learn how to use it.

You'll also need to learn how not to get stuck on single points. For example, in the prototype map, there are four boundary states which together define the complete field of possible discoveries—theory (theoretical science), the real world (real world science), the mind (mental science), and the body (physical science). Taken together, these states represent scientific genius. But taken individually, they merely compete with one another. Why? Because each of these four natural sciences is but a single point. They literally connect to nothing, have no direction, and begin where they end. Know it's this lack of connection which creates the problem. As I've said, it's the recognizable patterns of connections in things which give them meaning. Thus to have meaning, each of these scientific states must connect to another. Moreover, to produce scientific genius, they each must connect to the other three. And to see why, consider this.

Imagine you've been handed a picture of the Big Dipper. Here, someone has blackened out all but one star. Now ask yourself. Could you tell—with certainty—from this picture, which constellation this one star came from? Could you identify this star at all?

In truth, it wouldn't be possible. We assign stars much of their meaning from the patterns they make with nearby stars. And this is my point. If you isolate individual stars, they lose much of their meaning. Moreover, it turns out this holds true for all single points—celestial and otherwise. Connections are what breathe life into single points. Connections are what give things meaning.

In effect, these connections are what give stars their meaning. We call these patterns of connections, *constellations*. Moreover, this idea applies to all natural things. For instance, we call constellations of atoms, *molecules*. We call constellations of symptoms, *diseases*. We call constellations of national law, *constitutions*. And we call constellations of natural things, *genus*, *species*, and *family*.

Are you beginning to see the point of what I'm saying? If not, then try to focus on the last word in this sentence (the word, *sentence*). Now try to focus on this word for a whole minute *without moving your eyes*. (Don't scan the letters.) What you'll find is that your mind will go blank. And the longer you do this, the more blank your mind will become. Talk about not being able to access your whole mind. Moreover, once your mind goes blank, this word will lose all meaning. And herein lies one of the great dilemmas with regard to the current scientific method.

According to the current scientific method, all scientific endeavors must begin with a single point; a starting point. Scientists call this starting point, a *hypothesis*. The method then forbids these scientists from straying outside of this hypothesis. This forces scientists to deductively explore only this single point. Deduction means focusing inward on a single point.

Unfortunately, the longer a scientist—or anyone—focuses on a single point, the blanker their minds get. And the blanker their minds get, the more this point loses meaning. No surprise, scientists rarely discover anything meaningful—except by accident. Accidently, they expand the scope of their search beyond this single point—beyond their hypothesis.

As for the four single points I've been calling the four natural sciences, at this point, I recommend you act as if you've never heard of these words. Rest reassured, as we connect these points, their meanings will emerge. But only then will we have defined these words scientifically. And this is my point. *Unconnected single points have no meanings.* Connected points which make patterns do.

"Connections Between Single Points" IS What You Discover

When it comes to the scientific method then, this idea turns out to be a critical piece of information. Obviously, whenever you look to discover things, you must start somewhere. And this somewhere must be a single point. At the same time, since single points gain meanings only when they connect in patterns, if you fail to connect them to other points, you'll fail to discover their meaning.

For example, take obesity—something science currently struggles to understand. Obviously, so far, their search for an answer is not going well. People are fatter than ever and no one knows why. But were you to use the new method—constellated science—you'd discover a whole lot about these failures. Including how most of them stem from not connecting weight loss's critical pair of points—*eating less* and *exercising more*—to the attitudes people use to get themselves to eat less and exercise more—*will-powered force*.

What I'm saying is, in most cases, when people eat less and exercise more, they lose weight in the short run. But then they gain it all back and then some. The mystery is why this happens. Do you know why? It turns out that forcing yourself to eat less and exercise more is correlated to lasting weight *gain*, not loss. Indeed, most studies show that dieting—forcing yourself to eat less—almost never leads to lasting weight loss. There's a similar correlation between forcing yourself to exercise and not wanting to exercise. To wit, the more you force yourself to go to the gym, the less likely you'll be to keep going.

Where am I going with this? To see, watch what emerges when I connect these single points in a logically ordered, bulleted list. Moreover, if you pay close attention to the order of these points, you should see how they connect. Indeed, the more connections you see, the clearer these meanings will be.

So, why do people fail to lose weight?

- People are told they must *eat less* in order lose weight.
- People are told they must *exercise more* in order to lose weight.
- People are told that forcing themselves (using will power) to *eat less* is fine, and in fact necessary (no pain, no gain).
- People are told that forcing themselves (using will power) to *exercise more* is fine, and also necessary (ditto).
- For most people, forcing yourself (using will power) to *eat less* results in temporary weight loss—and long-term weight gain.

- For most people, forcing yourself (using will power) to *exercise more* results in temporary weight loss—and long-term weight gain.

Do any patterns jump out at you in this bulleted list? In truth, the problem with our efforts to lose weight—and the solution—is right there in front of you. To see it, just notice the pattern that surfaces within this bulleted list. The idea that *forcing yourself* to eat less and exercise more correlates to *weight gain over time*.

This, then, is a good example of how failing to connect single points can lead to problems. It also shows you how logically connecting single points can reveal these problems. So if you think back to what it felt like to read the opening paragraphs in this section—and if you then compare this to how it felt to read the bulleted list—you'll see what I mean. Reading the first three paragraphs in this subsection? It feels overwhelming. There are too many, unconnected single points. Whereas reading the points in the bulleted list causes patterns to emerge. Why? Because they're logically and visually connected.

Overall then, the thing to see here is how single points are like increment lines on a ruler. For a ruler increment line to have meaning, you must see how it connects to the other increment lines. Know this is true for all single points, whether seemingly big or small. To discover their meanings, you must connect them. No connections. No meanings.

How Serious Can It Be to Miss a Single Point?

Sometimes scientists make claims which, on the surface, appear to be good science. Often though, when the dust settles, you find they have overlooked a critical point. For example, take the St. John's Wort study undertaken by NCCAM (The National Center for Complementary and Alternative Medicine) in 2002. NCCAM is the federal government's lead agency for scientific research on complementary and alternative medicine. And in the April 10, 2002 issue of the Journal of the American Medical Association, NCCAM claimed that an extract of the herb St. John's Wort is no more effective than a placebo for treating major depression of moderate severity.

For the folks who love to fault herbal medicines, this may sound like a victory. Big Pharma just loves to rub these kinds of "scientific" surveys in the noses of more open minds. Unfortunately, the designers of this four year, six million dollar study failed to connect the dots. Moreover, in hindsight, these points are so obvious, it's amazing they have the balls to call this study, "scientific."

What did they overlook? For one thing, that doctors often tell people with depression to spend time each day under a sun lamp. Why? Because this increases the body's production of serotonin. Indeed, in a related study done in 2006, scientists compared the effects of light therapy and Prozac. The findings? The effect of light therapy was equal to the effect of Prozac—*and the light therapy helped more quickly.*

Can you guess the flaw in the first study yet? Well, if you read the list of possible side effects for St. John's Wort, you will. The main side effect of St. John's Wort is that, in large doses, it can cause severe reactions to sun exposure. But if you read the conditions of the NCCAM St. John's Wort study closely, nowhere do they mention sunlight as part of the test criteria.

Still don't see it? It's simple. If you take St. John's Wort and expose yourself to sunlight, the St. John's Wort raises your serotonin levels more than with simple exposure to sunlight. Whereas if you don't, St. John's Wort does little to nothing.

No surprise then that the NCCAM study found St. John's Wort has little to no clinically positive effect. Without exposing yourself to sunlight, the study is correct; St. John's Wort does next to nothing. Duh!

Here then is the constellation of single points I've just connected.

- People use St. John's Wort to treat depression.
- People use sunlamps to treat depression.
- In a 2002 study, St. John's Wort was found to be no better than placebo when compared to the SSRI, Zoloft (sertraline).
- In a 2006 study, light therapy (exposure to a sun lamp) was found to be quicker and equally effective to the SSRI, Prozac (fluoxetine).
- The main side effect of St. John's Wort is that, in large doses, it sensitizes you to sunlight.
- The 2002 St. John's Wort study failed to include exposure to sunlight as part of the test criteria.

Can Single Points Really Cause Personal Problems?

Now let's take a look at how this problem extends far beyond flawed scientific studies. Indeed, as it turns out, failing to connect single points is also the main cause of interpersonal disagreements. It's also the main reason people fail to resolve said arguments. And to see why, think back to the last painful argument you had.

Now try to remember where it all went wrong. Almost always, someone will have gotten stuck on a single point. For example, here are four of the more common complaints people raise in couples therapy.

"You never listen to me."
"You don't care how the kids misbehave."
"You always make your friends more important than me."
"Your boss doesn't give a shit about us."

Ever say—or hear—these things? Did they lead to an argument? Amazingly, if you'd *connected* a second point to the first, the argument would have fallen apart. But if you introduced this second point, then failed to connect it to the first, the argument would just have stalled.

For instance, consider the following complaint and retort.

"You never listen to me."
"I'm listening to you now."

Here things could go either way. If the person making the second point merely talks past the first person, then the argument will just get worse. But if both people hear both points and connect them, then a higher truth may emerge—that sometimes the second person does not listen. But sometimes he does and just does not agree.

Another example would be:

"You don't care when the kids misbehave."
"You never back me up when I do care."

Again, these two points could go either way. As single points, they could both be true. And felt separately, they would both feel painful. Connected, though, these two points could lead to a new understanding; that the second person does care, but has given up. And that both people—in their own way—feel equally alone.

Can Single Points Make People Blame?

Another thing to know about single point truths is that if you keep repeating them, they feel like blame. Indeed, were you to analyze most disagreements—scientific and otherwise—you'd find, this is what makes arguments painful. Ironically this holds true even when the points being made are about something good.

For example, say someone tells you you're smart. Obviously, this can feel good. But say this person then slowly repeats, "you're smart," five times in a row. Can you imagine someone saying this to you?

Even if spoken in a calm even voice, this can feel like they're telling you you're dumb. Of course, in some cases, people deliberately do this in an effort to batter someone into submission.

For example, I once saw a clip of Bill O'Reilly interviewing Michael Moore. In it, they both kept demanding the other respond to their point. Know the particular points they were each making are not important. What's important is that, in many ways, this interview resembled two boxers, repeatedly punching each other—but with words, not fists.

Sadly, the effect of these punches was not particularly salient nor relevant. Rather, because they each kept repeating their point, these two points quickly lost all meaning. The thing to see here is how repeating a single point can turn even the points of intelligent men into an assault on the other person. Perhaps that's why I can't remember their points.

Either way, if you want people to hear what you're saying, then remember to connect your points to theirs. Otherwise they may feel attacked, in which case your efforts will likely fail.

Success, Failure, and Connecting Single Points

Did you know that feelings of failure also come from not connecting single points? For instance, take *being* and *doing*. As I've said, *being* refers to seeing things change—and *doing* to seeing things not change. The thing is, to feel you've succeeded, you must see both these things at the same time. See only one and you'll feel like you've failed—at the part you don't see.

So say you focus only on *being*. Most times, you'll forget about your goals. Then—with no end in sight—you'll feel like you're always working, but fail to get anything done.

Focus only on *doing*, however, and you'll lose your ability to value what you do accomplish. You'll be so busy aiming, you'll fail to stop long enough to take proper credit.

For example, take how people feel when they go for a college degree. Often, after years of hard work, they graduate, then barely pause long enough to take it in. I've even had people tell me things like, "Yes, I have a master's degree. But I really need a doctorate."

So what should we call the six years it took to get that masters degree—a waste of time—a futile effort—something anyone could do?

Shouldn't this degree at least count as six successful years of hard work?

The solution? It's simple. Connect *being* to *doing*. As soon as you do, the problem gets solved. You also see where and how you've succeeded and what it took to get there.

In truth, no accomplishment feels real until you connect where you've come from to where you are now.

And if you fail to do this?

Then you'll end up feeling like you're always chasing rainbows.

The good news? How successful you feel is largely up to you. And if you remember to connect the dots, this will definitely change your life.

If Single Points Lead to Such Problems, Why Use Them?

Some may ask, why we use single points if they can cause so many problems. For one thing, we use them because all scientific endeavors must begin with single points. We also use them because most people hate not knowing things. So even negative points can feel better than none at all.

Then too, we use them because we love to come up with ways to refer to life's grand designs. So we use words like God and the Tao and forget they're each treating this grandness as if its a single point. At first glance, each of these words does appear to encompass a whole truth. However, no words can truly describe the depth and beauty inherent in words like God and the Tao.

No coincidence, the Tao itself begins with cautionary words to this effect. "The Tao that can be told is not the eternal Tao." Here, the Tao is saying, profound truths cannot be put into words. And when people use words like God casually, it's easy to see this as true.

My point is, at best, these words vaguely refer to these sacred spiritual beliefs. So if you fail to treat them as single points, you may end up believing you can actually comprehend the ineffable.

Yet another thing which makes us gravitate to single points is that, by nature, we're programmed to avoid suffering. And in theory, being able to predict the next suffering can give us the power to avoid it. Indeed, were you to survey the nature of scientific endeavors, you'd find this goal at the heart of many. No coincidence, in a future chapter, we'll look at the idea that most of science is driven by the desire to avoid suffering.

The problem is, as I've said, unchanging things can exist only in theory. Thus single points—such as an effortless diet or the perfect teaching method—can't exist in the real world. Why? Because in the real world, single points are like the ends of a rainbow—they're beautiful but unreachable goals. In other words, they're moving targets which, like all moving targets, are notoriously hard to hit.

Then there's our obsession with winning arguments. Admittedly, it can feel good. And if we hammer someone with a single point, we can often make our opponent lose their words—and give in. Unfortunately, as with all unconnected single points, any satisfaction we feel will be short lived. In other words, for an argument to feel satisfying, it must create connections. No connections—no discoveries, and no discoveries—no satisfaction.

This is why—even when you win an argument—the good feelings rarely last. This makes winning arguments yet another thing which resembles trying to reach the end of a rainbow. In theory, a single point can let you win an argument. Certainly, it can end one. But in the real world—like being done, satisfaction is always a moving target. And this is where the second geometry comes in.

Linear Continuums
(the second geometry: the Measuring Tool)
© 2008, Steven Paglierani, The Center for Emergence

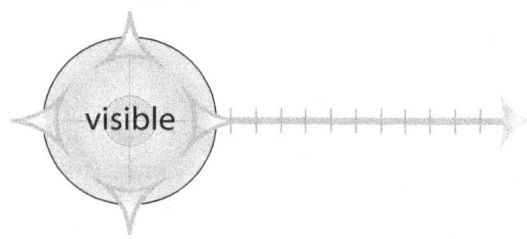

a single point,
with one direction and no connections

Linear Continuums (the Second Logical Geometry)

Okay. If you don't connect single points to other points, all manner of problems arise. If you keep this in mind though, you can avoid things which would otherwise prevent you from making discoveries. Know however, when you get stuck on a single point, if you can move, there is a way out. This is where the second geometry comes in. The second geometry enables you to move toward another single point.

Please realize that without the second geometry, there would be no science. It alone describes the act of measurement—and measurement is an integral part of all science. But while this description may make the second geometry seem straight forward, in truth, it's anything but. Indeed, when you look beneath the surface, you find, there's a lot going on.

What's going on? To begin with, let's compare the second geometry to the first. Single points exist only in theory. Linear continuums exist partly in theory—and partly in the real world. Here the theoretical part is the starting point—the part which is unchanging. And the real world part is the moving end—the part which is constantly changing.

Know it's the constantly-changing part which holds the key to understanding the second geometry. To see why, imagine you need to measure the width of a wall. Consider how you'd begin. You'd likely have someone help you, by holding one end of a measuring tape against one end of this wall. You'd then extend this measuring tape toward the other end.

Know it's this movement—away from one point towards another—which creates the linear continuum. In effect, a linear continuum is a single point dividing into two. Here, one end remains still while the other starts moving. And this movement allows you to measure the space between two points.

Realize, the moving end of this continuum is not the second single point. Why? Because the two points I'm referring to are the two ends of this wall. Moreover, something important happens the minute you read the measurement. The movement stops and the continuum collapses.

This then transforms the entire second geometry into a theoretical object—a result—a single point. Thus a second geometry exists only for as long as a measurement is "in progress." Two examples of this would be the ideas of scientific and technological progress. The single point they're moving toward? An ideal world, wherein we're all happy. The thing is, as soon as they reach the next level, meaning, the end of the current wall, both science and technology collapse back to a current, single point.

What can we learn from this? That while newer sciences and technologies may incrementally improve our lives, at the same time, there is always more to discover. Understandably, many folks find this truth hard to accept. We'd like to believe we can discover ways to make our dreams come true. But dreams are also always single points,. Thus dreams exist only in theory.

At the same time,the second geometry explains why being unable to reach our dreams is never failure. Failure is remaining stuck on a single point—and measuring is moving. Indeed, the desire to reach our dreams is largely what makes single points begin to move. And each time we measure our progress, it feels like we've moved closer to this goal.

The problem of course is that each time we point to this progress, we risk reducing our progress to a single point. This is why, when we celebrate progress, the good feelings are often short-lived.

This also explains how we can fall prey to false beliefs like that "more is better." This belief is rooted in the ambivalence which exists between the first two geometries. For example, we sometimes think that if a dessert is good, that we should have more—more is better. Or that if sex feels great, then more sex would be better. Or that the old days were better. In theory, these ideas are true. But in the real world—where all things keep changing, good feelings always change into fullness, soreness, and longing for the past.

So what do we gain from the second geometry? We gain a sense of direction and purpose. We also see proof for our progress—which can then keep us from giving up. We also get a way to keep our dreams alive as we pursue our dreams. The thing is, when it comes to a scientific method, we need more than just direction, theoretical measurements, and a sense of progress. We also need it to make real world measurements, and useful discoveries. What kinds of discoveries are useful? The kind which reveal the patterns which exist between pairs of single points. Enter the third geometry.

Fractal Continuums
(the third geometry: the Possibility Defining Tool)
© 2008, Steven Paglierani, The Center for Emergence

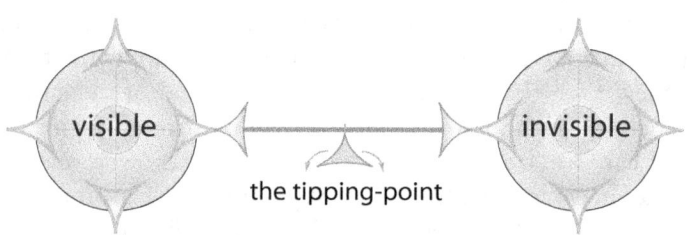

two single points,
with two directions and one connection

Fractal Continuums (the Third Logical Geometry)

Okay. From the first geometry we learn that all discoveries must begin with a single point. And from the second geometry we learn we must make measurements in order to advance toward discovery. The thing is, how do we choose a direction and how many measurements are enough? In other words, how do we limit our efforts to what's relevant—while leaving nothing out?

The third geometry addresses all these concerns and more. To do this, it uses a pair of single points joined by a thread of similarity. Here the two points are complementary opposites, in that they contain none of each other. Yet together, they bookend all things which exist between these two points. Moreover, when we connect these two points with what they have in common, we create a continuum. This continuum gives us access to everything between these points, allowing us to scientifically define, explore, contrast, and measure all things.

Can Science's "Real World Measurement" Problem Be Solved?

Arguably, it's the third geometry that opens the door to the new scientific method. Without this geometry, this method could not exist. Now to see what makes this geometry special, consider this continuum's two end points. Between them lie all instances of the first of two qualities all things in our world share. Here, I'm referring to the quality of visibility.

At the same time, despite encompassing an infinitely dense amount of content, we can still measure this content in the real world with 100% certainty. Either you can see a thing. Or you cannot. If you can, it's visible. But if you see nothing, then it's not.

Contrast these two outcomes to those of the second geometry, wherein content is measured by linear scales. Scales are straight lines with evenly spaced rules. Yet we learned in fifth grade that there are no straight lines which exist in nature. So why does science constantly claim it can measure natural things with straight lines?

This obsession with linearity then leads to some rather amusing flaws. And some not so amusing. For instance, take the present method's attempts to measure personality. Nowhere does science pretend to know more than when measuring who people are, and how they think and feel. Think about it though. Depressed? Angry? Unmotivated? Narcissistic? How can straight lines measure these things?

In truth, one person's behavior (what you can see) and inner life (what you cannot see) often differs markedly from another's. And even a single person's behavior and life differs one minute to the next. Now consider how hard it must be to measure multiple people over time. Surely, they can see these problems.

Rather than admit these limitations though, science turns to magical, made-up measures. Here, almost all of them are based on linear scales. This means they can be true *only in theory*. To overcome this problem, they use the same method weatherwomen and weathermen use. They use words like "partly" and "chance of" to make their statistical outcomes look real.

What about medical prognosis? Can we really assign numbers to people's chances? And how do these vague insinuations affect people's chances to recover? In truth, we can't make shoes or shirts or dresses where the sizes match—let alone predict recovery from an illness or injury.

Please know, I do not blame scientists for any of this—seriously, I do not. We pressure doctors to tell us our chances. We blindly trust our weatherwomen and weathermen. Moreover, we expect talk therapists—who are only people, after all—to find, improve, and save our relationships. Yet the last time I checked, science has yet to scientifically define what a wound is, let alone things like love.

Having accurate, real world measurements would change all of this. You can't change what you can't see, and accurate measurements would help us to see. Fortunately, the third geometry allows us to make these kinds of measurements, in effect, giving us a direction. And to see this in action, let's explore a real world problem—closed mindedness. Here, I challenge anyone—scientist or otherwise, to come up with a measurable, empirical, scientifically-sound method for opening minds. I'm serious. In fact, before you move to the next subhead, please take the time to actually try to do this. At the very least, try to come up with a short list of what has prevented science from doing this. What methods have they tried to use?

Can the Third Geometry Open a Closed Mind?

When most people talk about something important, they end up focusing on a single point. Most listeners then either agree or disagree with this point. The thing is, conversations which focus on single points close minds. And this is true, even when the point is something positive.

For example, imagine someone tells you, "I love you"; obviously, hearing this can feel nice. Now imagine this person says this to you, over and over again. Because our brains are programmed to favor novelty over repetition, when things repeat, we rapidly lose touch with what's being said. In effect, this repetition dulls our senses—and these nice words now feel hollow.

Interestingly enough, introducing the opposite point can at times remedy this problem—especially if the two points get placed at opposite ends of a continuum. However, this is only true if both points stay in plain view. If they don't, then a new argument will likely ensue over which point is more relevant.

For example, take the repetition we just spoke about—being told, "I love you." Now imagine hearing someone repeatedly saying this to you. Now clear your mind and imagine someone telling you, sometimes they love you and sometimes they just want to kill you. Given the "kill" part isn't literal, most people experience more good in the second version.

Why does juxtaposing good and bad things feel better to us? Because even good single points shut people down. Fortunately for us, the remedy is easy and simple. Introduce an opposite point, then juxtapose it to the first. This engages people and makes things personal and real for them.

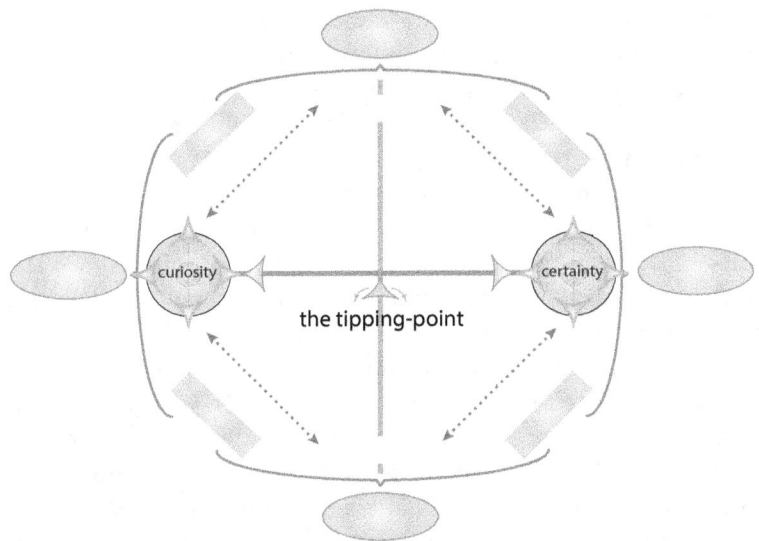

Constellated Science's
Defining the Open / Closed Mind
(the gateway to all discoveries)

(undefined) (vertical axis question)
Is this mind open? (horizontal axis question)

Can Science Actually Measure the Openness of a Mind?

Admittedly, some may dismiss what I've said as simply being clever. Worse yet, you may fail to see the science I've used in this example. Fair enough. In truth, you're right to question my claim—that I've defined a way to open minds. After all, this openness is the requisite doorway into everything from talk therapy and physics to education and medicine.

Obviously, to do these things, people's minds must be open. But to prove my claim, we must have a way to define and measure this openness. So how do I define an open mind? Here again, we'll need to turn to the third geometry. With it, we can define our terms, and also design real world measurements which test open-mindedness. Let's start with the phrase, an "open mind." How do we scientifically define an open mind?

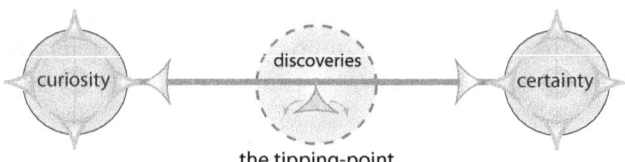

How Do I Define An Open Mind?

To begin with, we'll need to identify this point's complementary opposite. Fortunately, doing this is easy. The complement to an open mind is a "closed mind." Easy enough. The problem, of course, is that the words "open" and "closed" do not scientifically define anything. And as I said, to proceed, we must define our terms.

At the same time, I don't want to get ahead of myself by creating a whole map. We'll do that in future chapters, when we map the essential qualities of science, therapy, learning, and relationships. It won't hurt though, to use a blank map to define our terms. After all, we'll need these definitions when we make those other maps.

Okay. Can you see what I've done in this drawing? I've used the prototype map's horizontal question—*does this thing change?*—to create a fractal continuum. Here I've chosen the word "curiosity" to represent the open mind—and "certainty" to represent the closed mind. As to why these two words, again, know we'll talk at length about this in future chapters.

For now, let's just say these two words represent the sine qua non of these two states of mind; the essence of these two qualities. What I do need to tell you now though is how I came to place these two words where they are. To do this, I looked for parallels between these two words and the placements of the qualities in the prototype map.

To wit, I asked myself, do the qualities *visible* and *invisible* apply to these two words? And they don't. I then asked myself if the qualities *changing* and *unchanging* apply? And they do. In other words, I asked myself if either curiosity or certainty is an unchanging quality? Sure enough, one word is and one is not. Certainty is any belief that requires no change to be true. Conversely, curiosity is any belief that includes the possibility of change in its truth.

Can you see yet what makes me say these terms are complementary opposites? No certainty exists in curiosity—no curiosity exists in certainty. Together, though, they comprise all possible reactions to suggestions involving change. And being open to all possibilities meets our goal.

Can you also see how this pair of opposites allows us to make accurate, real world measures? Either your beliefs are changing or they're not. If they are, then you're curious—and your mind is open. If they're not, then your mind is certain. And if you're certain, your mind is closed.

What about times wherein you're almost certain? Don't these times qualify as a flaw in our ability to measure openness? Actually, if you think about it, they don't. If you're considering changing your mind, then you're open to change and so, not certain. Can you also see how posing things as complementary opposites makes defining them easy and clear? Curiosity becomes the state of mind wherein your beliefs are changing. And certainty becomes the state of mind wherein they're not. And even 5% change must still be considered change.

So now, compare what I've just shown you to the four qualities I mentioned in the beginning of the previous chapter. There I said a real scientific method must include four qualities.

- **The Process Must Contain No Ambiguity**: these questions—and all terms used—will be clearly defined.
- **The Process Must Be Unbiased**: these questions—and all terms used—will always account for all possibilities.
- **The Process Must Be Pragmatically Testable**: these questions—and all terms used—will lead to practical, real world tests and yield accurate, real world answers.
- **The Process Must Employ No Statistics**: the results of all terms and tests will be 100% certain.

Does our partially finished map of the open / closed mind meet these standards? Without a doubt, it does. And this raises an important question. When I say "without a doubt," is my mind closed? My answer may surprise you. My answer is "yes." And "no." How can this be? All I'll say for now is, the answer lies in how fractals are simultaneously linear and nonlinear. Meaning what? This may take a few sections. But here goes.

Can Fractals & Nonlinearity Be the Key to it All?

If you remember—in the reader's guide—I defined the word, *discovery*. There I said *discovery* is *the act of connecting single points in meaningful patterns*. Sounds simple. At this point though, hopefully, it's becoming clear that I have yet to define a critical word in that definition. To wit, I have yet to define the word, *meaningful*.

So what do I mean by *meaningful*? I mean, this thing has two qualities—it's nonlinear and it's fractal. Let's start with the nonlinear part.

To begin with, according to the prototype map, the horizontal axis always refers to the quality of change. And at the left end of this axis, what we're exploring does change—and at the right end, it does not. In between these two points then, things change unevenly—as they transition between theory and the real world and back. And the word science uses to refer to this kind of change is *nonlinear change*.

Know that *nonlinear change* happens to be one of the main differences between the third geometry and the second. And to see this difference, consider how clouds change as they move across the sky. Now consider the way measurements change as you move your finger along a ruler's edge. Can you describe the difference?

Okay. It's hard. But I'm sure you can tell me what these two changes look like. As clouds move, they change in uneven, mysterious ways. As your finger traces along the edge of a ruler, things change incrementally. Here, *incremental* refers to the idea that rulers change in evenly spaced increments—in a linear fashion. As opposed to clouds which change in a nonlinear fashion.

The big things to see this time involve how we feel about linear change. It turns out, we humans love it. To us, doing things in a linear way seems simpler, and easier to understand. Also, humans tend to look for the shortest, simplest ways of doing things. Quicker and easier is better, yes? Oh were it only possible.

The thing is, I've already mentioned there are no straight lines in the real world. In the real world, nonlinearity is the norm. So much for the dream of the easier softer way. It simply does not exist. But we all, including science, keep pretending, real world linearity exists.

In truth, this tendency turns out to be one of the fatal flaws in the current scientific method. Here, scientists are taught to see linearity as the proof something is true. And in theory, they're right. This is why scientists try so hard to get experimental outcomes to turn out the same every time. Measurements that turn out the same every time are linear measurements.

Ironically, in the real world, linearity is the proof something is *not* real, as well as that this thing is not true. In the real world, where there are no straight lines, nothing ever repeats the same way twice. In other words, everything's nonlinear.

Finally, remember, I've already told you that the horizontal axis isn't linear—let alone incremental. Yet most people still imagine this space as being evenly distributed along the whole line. Like a ruler. It's not. But in order to see why, you have to grasp the idea of *fractals*. And as you're about to see, fractals are the closest nature comes to linearity.

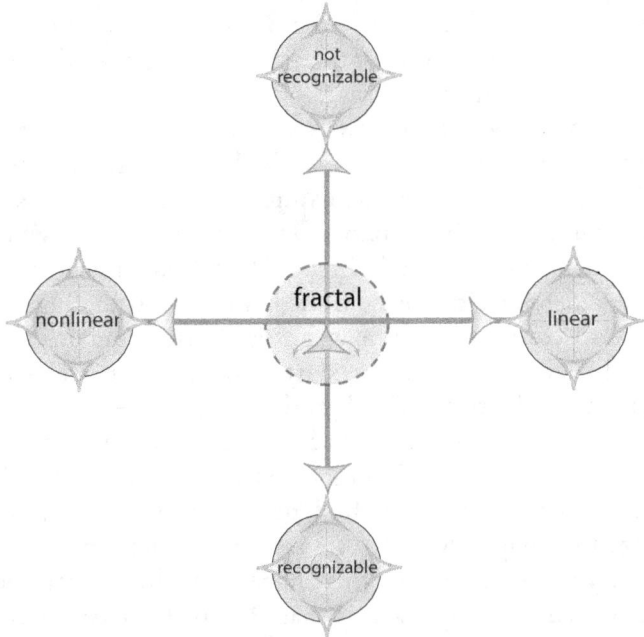

Constellated Science's
Mapping the Word "Fractal"
(a necessary quality in all discoveries)

Is this pattern recognizable? (vertical axis question)
Does this pattern change? (horizontal axis question)

What Are *Fractals* Anyway?

Now let's talk about how being fractal makes things meaningful. What are fractals anyway? In 1975, a brilliant IBM researcher named Benoit Mandelbrot coined the word "fractal." He made up this word to refer to a type of nonlinear geometry—things like coastlines, and cloud shapes, and crystal formations.

What makes this geometry meaningful? Self-similar at different scales. And if I just lost you, don't feel bad. This concept is so complex, whole books have been written trying to explain this term.

In the late 1980's, I read one of these books—James Gleick's book *Chaos*. His description of fractals is what started me questioning the present scientific method. Then, in part, fueled by my desire to share this personal discovery with others, I spent almost two decades trying to define this term. Finally, years later, the third geometry came to my aid.

So how do I define a fractal? A fractal is *a recognizable pattern which always repeats differently*. As opposed to a linear pattern, which is *a recognizable pattern which always repeats identically*. Can you see the pair of complementary opposites here—the words *differently* and *identically?* Here, each word contains none of the other. Yet together, they define all the ways we can recognize patterns.

At the same time, while these two words are complementary opposites, the rest of these two definitions are exactly the same. And together, these two definitions contain a balanced mixture of change and no change. And it's this dual nature—the linear part and the nonlinear part—which holds the key to understanding fractals. Fractals have both parts in one pattern.

Not coincidentally, fractals hold the key to understanding the nature of scientific discoveries. All scientific discoveries contain a balanced mixture of change and no change—part theory and part real world. And to see what I mean by this, go back to the image I referred to a moment ago—the image of a cloud moving across the sky. Let's say this cloud is one of those puffy ones—the kind scientists call "cumulus clouds."

Now ask yourself this. As a child, how long did it take you to learn to recognize this shape? Now ask yourself how much time you've spent since then trying to remember this shape? Finally, ask yourself if you could you ever fail to recognize this shape? In truth, once you've learned to recognize this pattern, you know this pattern forever.

Are All Discoveries, *Fractal?*

What I've just said is, cumulus clouds are *a recognizable pattern which always repeats differently*. Know this is what makes clouds, fractals—in particular, the "recognizable" and "repeats differently" parts. In effect, while no two cumulus clouds will ever be the same—and while no single cumulus cloud will ever form the same shape twice—still, without effort, even young children can learn to recognize and retain this shape.

Know it's the recognizable part which I'm calling the linear part—the part of these clouds which does not change. And the rest is what I'm calling the nonlinear part—the part which never repeats the same way twice. And yes, I realize, it may sound strange for me to call something without straight lines, *linear*. My answer? Go back to my definition for linear patterns—they're *recognizable patterns which always repeat identically*.

Now, did you notice that nowhere in this definition do I mention straight lines? To our minds then, although none of the lines in these clouds repeat the same way, in another sense, these clouds are identical. Indeed, it's this unchanging recognizability that accounts for the linear part of fractals. Scientists even have a name for unchanging recognizability. They call it *self-similarity*.

I call self-similarity, "nature's linearity." At the same time, this recognizable pattern is the only part of these clouds that repeats the same way. The rest of these clouds remains forever nonlinear. And this combination is what makes fractals—and the third geometry—so amazing. And meaningful. Half of every fractal is recognizable—and half is not.

In addition, since fractals never repeat the same way twice, each time you recognize a fractal, you make a new discovery. You discover this pattern is one you've seen before—just not in this form. Know it's discovering this connection which underlies all scientific discoveries. This connection reveals yet another of nature's universal patterns. And since all fractal patterns emerge from the same six shapes—the six logical geometries—these six geometries are the essence of all scientific discovery.

The Third Geometry: the Nerd's Description

Some may have noticed the three qualities I've used to describe the previous two geometries—[1] the number of points, [2] the number of directions, and [3] the number of connections. Know I'll continue to use this format for the remaining geometries. Why? Because it's important to see how these complex shapes relate—as well as how complex they are.

The third geometry? It's a two-point set, with two directions, and one connection. Know this simple description barely hints at the complexity hidden here. For instance, take its starting points—fractal continuums have an infinite number. As long as you stay on the continuum, you literally can start anywhere you want.

They also have two forward directions (left and right) and no backwards directions. In effect, all movement is forward. At the same time, in theory, they have two end points. But since the meanings of these two points stem entirely from the connections they make, even these end points are starting points as well. Did I just lose you again? If so, know this is not your fault. All expressions of the third geometry contain an infinite number of things to discover—clearly a lot to take in.

Can Fractal Continuums Open Minds?: Part 2

Now let's take another look at how the third geometry opens minds. For example, take the idea that some people claim anger is bad, in effect, that we shouldn't get angry. Indeed, lots of therapists and spiritual teachers make this claim. So are they right? Should we do all we can to avoid getting angry?

To see for yourself, place this idea—that anger is bad—on one end of a fractal continuum. On the other end, posit an opposite idea, such as that therapists often do everything they can to help people to get in touch with their anger. Finally, connect these two ideas—that anger is bad and that anger is good. In doing this, you create the mental conflict necessary to provoke a potential healing crisis.

For instance, say someone tells you that your anger is bad. Okay. Ask them to consider what life would be like if you were unable to get angry. This stuckness happens to a lot of people. But in all likelihood, the person asking you has never considered this side of the coin. You might then ask, "Do you realize how holding in anger can lead to getting walked on." Or how being unable to get angry can lead to depression or rage? At which point, hopefully, these counterpoints would pleasantly surprise the person. Perhaps, it might even provoke a healing crisis.

A second example would be people who complain that their mothers are controlling and overinvolved. Here again, juxtaposing this idea with its mirror image can often open people's minds. For instance, I once asked a woman who felt this to imagine having a mother who was cold, distant, and uninvolved—this surprised her and caused her to pause. I then asked her how involved in her life her father was. Sure enough, she described him as uninvolved.

Can you see the pair of complementary opposites yet? In a later chapter, we'll look at how most relationships contain many of these pairs. In each case, juxtaposing the pair holds the key to healing childhood wounds. And in this woman's case, I created a fractal continuum containing her parents' involvement with her. Then, when she saw it, she realized she'd been blaming everything on her overinvolved mother. What she hadn't realized was what it must have been like for her mother to parent alone.

These kinds of previously unseen ideas about parents frequently emerge when you employ fractal continuums—especially when one parent has been overinvolved. Or unemotional. Or too angry. Here, many people get stuck on a single point—such as that the overinvolved parent was the bad parent. Or that the unemotional parent didn't love them. But when you juxtapose these aspects of parenting on a fractal continuum, people often come to see their parents in a new light. For instance, they may come to see that having an under involved parent hurts just as much, or that it's the unemotional parent who is cool and calm in a crisis.

A third example involves me and the way I used to feel about getting on scales. I hated weighing myself—but felt compelled to do it anyway. Long story short, I used a technique I'll describe in a future chapter— Cycles of Three. I used this technique to connect my feelings about getting on scales to two scenes from my early childhood.

In one, I felt scared I'd die from being too fat—the world's standard message to overweight people. But in the other, I felt scared I would die from being too thin—my mother died of anorexia.

Can you see the continuum here? Both these things had felt true. And when I realized this, I saw the absurdity of trying to choose between these points. In that moment, years of fears melted away and I laughed. Yet another example of how fractal continuums can heal childhood injuries—by revealing previously unseen choices.

Finally there is the idea that these previously unseen choices are what you discover. This, in part, is why discovering things makes you feel so good. You feel good because—by defining your search—you discover pain-free options. This frees you from your previous limits, by revealing new possibilities. Hence the name, the "possibility defining" tool.

The Fourth Logical Geometry: Tritinuums

So far you've identified a starting point (the first geometry), taken measured steps out from this starting point (the second geometry), and connected this starting point to its opposite (the third geometry). You've defined the limits of what you're about to explore. Now you must connect yourself to what you're seeing. Thus despite warnings to remain impartially detached, in truth, scientists must personally connect to whatever they explore. Connections are what make us conscious.

Admittedly, grasping why this is so can be somewhat hard. To understand, you must understand tritinuums. Everything we do to make ourselves conscious involves a tritinuum. And while this may make them sound grand, they're really not. Especially if you keep in mind that tritinuums are just triangles of fractal continuums—and all we use them for is to triangulate on life.

Okay. I admit. This tritinuum diagram is the worst ever. If you take it slow though, I'm sure you'll do fine. You'll need to begin by doing something completely counterintuitive. Don't look at the diagram yet. If you do, in all likelihood, you'll feel completely overwhelmed.

Okay. You're not looking, right? Okay. Then we can begin. The first thing to know is what this diagram represents. It represents a rationalist scientist exploring puzzle pieces. Here, rationalist scientists look for two qualities in things—invisibility and unchange. And before you jump ship, remember, all we're talking about here is ideas. Constellated science defines ideas as invisible unchange.

Now look over at the drawing—briefly—and try to find the tritinuums. Hint. They're shaped like triangles. Can you see them? There are two—an upper and a lower. Here the lower one contains what this scientist is sensing out in the world—while the upper one is the mental mirror image of the lower one.

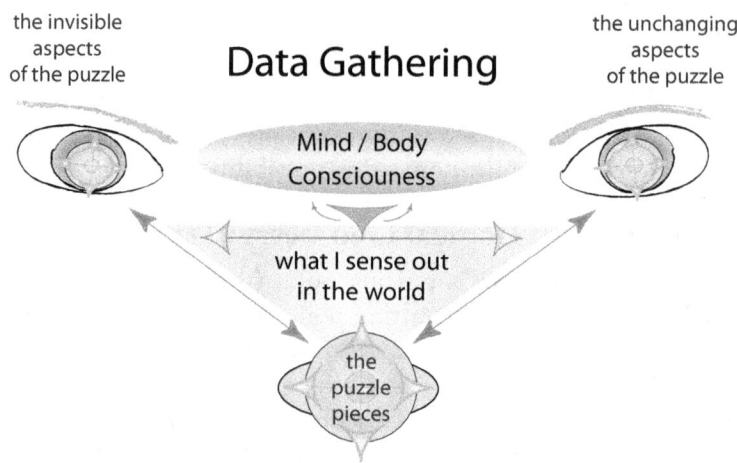

A Rationalist Scientist Gathering Data

So what is this scientist doing? He's about to start a 1,000 piece puzzle. I'm using puzzles because solving puzzles are a wonderful analogy for doing science. So where would you begin? If you're a rationalist scientist, you'd base your strategy on logic and reason. In other words, you'd look for the corner pieces.

Now look over at the drawing again and this time, look for the scientist's eyes. Now locate the puzzle pieces and draw the tritinuum in your mind. Together, these three points create the sensing tritinuum—two eyes gathering data from a third point. This is the data gathering part of the becoming-conscious process.

I realize, when we normally think of corners, we mainly think of things with right angles—mostly squares and rectangles. Tritinuums also have corners. But they obviously only have three. And in every tritinuum, two corners act like eyes, while the third corner is what these two eyes focus on.

Know that constellated science puts special value on corner pieces. Corner pieces are good starting points for just about any problem. Having three or more related starting points (*first geometry*) creates a well-defined space to explore. Having four corners is ideal. I'll explain why later.

Can you see the three fractal continuums which make up this tritinuum? They're the three double headed arrows. Now read the definition—"three single points, six directions, and three connections." At this point, you should be easily able to recognize all three parts—the three points (two eyes and one data point), the six directions (the six arrow heads, each going in a direction), and the three connections (the three arrows, each of which joins two points).

Tritinuum of the Rationalist Scientist Becoming Conscious
[tri · **tin** · u · ums]
(an example of the fourth geometry: the Consciousness Creating Tool)
© 2008, Steven Paglierani, The Center for Emergence

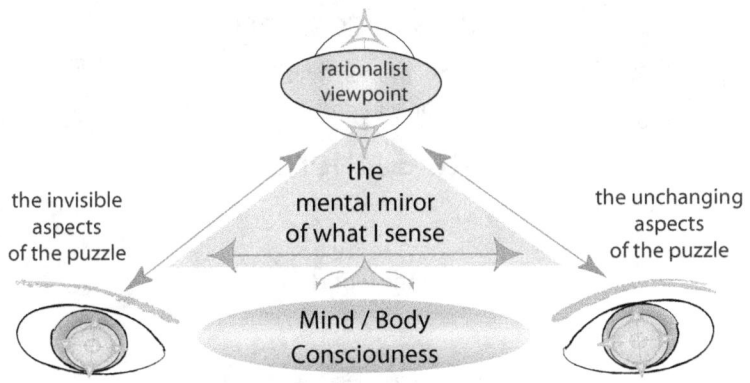

Data Processing

three single points,
with six directions and three connections

A Rationalist Scientist Processing Data

Okay. Take a breath. We're almost done. All we have left is to process our data. By this, I mean we need to look for patterns in the data we've collected. To do this, use the same two eyes. But this time, our eyes look inward—this time we're reviewing the data on the screen of our mind.

Realize this second half of becoming conscious is the complement to the first. And together, we have yet another baseline pair of complementary opposites. Of course, between this pair lies the goal of the whole endeavor—a pool of scientific consciousness. Moreover, everything we take in—and everything we process—passes through this pool.

Tritinuum of the Rationalist Scientist Becoming Conscious
[tri · **tin** · u · ums]
(an example of the fourth geometry: the Consciousness Creating Tool)
© 2008, Steven Paglierani, The Center for Emergence

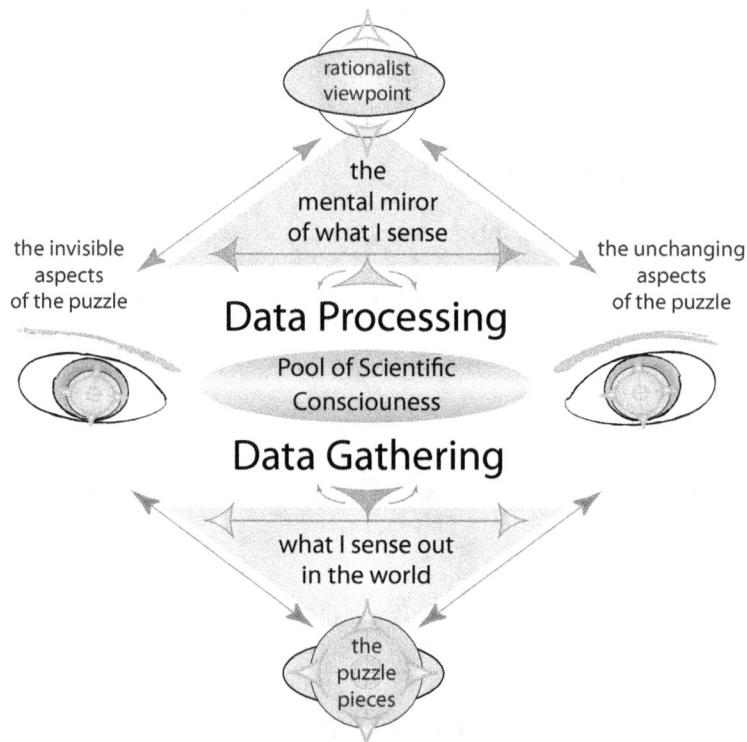

A Rationalist Scientist Solving a Puzzle

So now, say you've chosen to start looking for the puzzle's corner pieces. Obviously, you'd need to scan the pieces to find these corners (scanning is the second geometry). You'd then need to feed what you'd seen back to your rational mind.

Now look once more at what's going on in the data gathering tritinuum. In this moment, this person is looking for pieces with straight lines (*linearity*) and without (*the lack of linearity*). And as I said earlier, this juxtaposition of *linearity* and *nonlinearity* is the essence of all things fractal. Here, the invisible aspects equate to the nonlinear pieces—and the unchanging aspects to the linear pieces.

Admittedly, this last analogy can be hard to understand. Where did I even get these two qualities? If you go back to the prototype map though—and if you look at the four classes of scientific discoveries (material, empirical, rational, and spiritual discoveries)—you'll see they each derive from two adjacent aspects of the mind.

Literally, according to the map, each scientific viewpoint gets created by combining adjacent answers. Realize, it's two of these adjacent answers that I'm equating to the rational mind's two eyes. And the job of these two eyes is to understand the *invisible* and *unchanging* aspects of what we see. Moreover, this same analogy applies to the other three scientific realms as well.

For instance, the task of the spiritual mind's two eyes is to understand the *invisible* and *changing* aspects of what we see. One eye looks for invisibility; the other, for change. And if you look closely at the map of scientific genius, you'll see that a meta-thread of similarity connects each pair of eyes to each other pair.

With the rational and spiritual minds, this thread is *invisibility*—both minds have this in common. Moreover, it's our awareness of this commonality which allows us to become personally conscious of their differences. By doing this, we discover the sameness and differences between these two viewpoints. Including that there's only one main difference between Spiritual things and Rational things. Spiritual things constantly change. Rational things are unchanging.

Imagine? This is it. This is what they fight about.

A Big Question: Where Do Sensations Come From?

Okay. So take a breath and perhaps, now would be a good time to take a break. In part, I say this as I realize I've skipped something complex and important. This thing? I've failed to explain to you how tritinuums make doing science personal. Admittedly, explaining this may take a bit of time.

To begin with, all human sensations are rooted in tritinuums. It's tritinuums which allow us to gather—and collect—our impressions of our world. They also enable us to sort through, and process, these sensations.

Know we do this until we either discover meaningful patterns—or until we get lost and give up.

In effect, all sensations and indeed, human consciousness itself, begins with some version of this geometry. In a way then, we could say that tritinuums are the geometry of life itself. I say this because it's sensation which creates a bridge between us and our world. Here sensation is a sort of bidirectional feedback loop which allows us to perceive—and understand—our bodies, our minds, and our world.

For example take eyesight, a classic tritinuum—two points reaching out to explore a third. Here, each time we look out at something, our two eyes briefly land and focus on this third point. And if they don't land, then we feel disconnected and our minds go blank.

Conversely, if we linger too long, we also go blank, at which point this thing becomes invisible to us. And in both cases, this bidirectional feedback loop breaks, leaving us unable to see what's right in front of us—we either cycle between moving and pausing, or we end up temporarily blind.

Want to see how this works for yourself? Look out at the world—now scan what's right in front of you. Now continue scanning and don't allow your eyes land on anything. In effect, keep looking but don't stop. In no time at all, you'll feel blank, blind, and really confused—as your visual, bidirectional feedback loop collapses.

Know the same thing will happen if you remain focused on anything. And to see this in actions, just pick any object and keep looking at it for a full minute—*without moving your eyes, blinking, or looking away*. Again, you'll find yourself blank, blind, and confused, at least until you rescan this object. At which point, you'll restore this bidirectional feedback loop.

Know the main thing to take away here is the meaning of the word *sensation*. This word is short for "sensing change." If you perceive no change—you feel no sensation. Which is why we constantly move our eyes over still objects. In effect, in order to perceive anything, we must alternate between sensing it and pausing to understand. Tritinuums enable us to do both of these things. Remember too that our underlying goal here is always to find patterns in what we sense. This ability is what we have come to call IQ, yet another thing which stems from tritinuums.

Of course, if we can't find a pattern, eventually, we go blank and this sensing<>understanding cycle stalls. In these times, we literally lose our ability to connect to our world, and for the most part we become unconscious.

This then is what the fourth geometry gives us. It gives us the ability to collect, compare, and contrast what we sense. This sensing<>understanding cycle is what enables us to look for, and recognize, patterns—in ourselves and in our world. It also anchors our perspective, allowing us to become conscious of how we make these connections. And this awareness—and the connections it creates—is what we call, personal consciousness.

Are All Five Senses Based on Tritinuums?

Now let's take a deeper look at the bidirectional feedback loop we call sensation. We'll begin with how we gather sensations. Here, all sensations involve a pair of sensory viewing ports. This is why we have two eyes, two ears, two nostrils, two sides of the tongue, two hands, and two feet.

Obviously we also have two kidneys, two knees, a body with two lungs, a brain with two halves, and a butt with two cheeks. So what does all this bilateral physiology have to do with tritinuums? To begin with, in its simplest form, a tritinuum is just two points exploring a third. Here each of these pairs contributes two points to the tritinuum—the exploring points. The third tritinuum point then emerges when the mind and body pick a point to focus on. At which point, the loop gets established.

What makes this particular geometry so important? Consider what happens when a person loses an eye. In effect, their visual tritinuum collapses. And yes, in most cases, the person can still see and that's a good thing. At the same time, he or she loses his or her sense of depth perception.

This then is an important thing to know about tritinuums—they give us our sense of depth perception. Thus, to have any sense of depth perception, all three tritinuum points must be functioning. And if we lose one? To what this is like, just cover one eye and try judging distance.

Did you try it? If you did you know you could still see things. But you could see them only as points on a flat plane. To locate things in three-dimensional space, you need at least three points. You had two—the "x" and "y" axes (up & down, left & right). But you lost your sense of the third point; the "z" axis (depth).

Know I once came across an example of this when I was visiting an artist friend. That day, I noticed his cat seemed to move about rather oddly. He'd try to leap somewhere. But then he'd constantly misjudge distance or height. Then he'd bump his head, then stop for a moment, then walk away.

Now before you misinterpret my intention and laugh, consider this. If you were a cat, what would being like this do to your life? Can you imagine the hell this poor cat endured? His nature told him he could jump and leap—and that he was agile and fit. But since he lacked the ability to sense "z space," he'd constantly bump, bang, and hit into things.

Why bring this up? Clearly, it's a good example of how tritinuums do or do not create our senses. It also shows what happens when a sensory tritinuum collapses. Know that when I left that day, I'd managed to help my friend's cat get back to normal. How? To begin with, I noticed that when the cat moved slowly, he'd do fine. But when he tried to move quickly, he'd bump into things.

Know it's far too early to go into what I did that day. I promise to return to this story in a future chapter. For now, there are two things to see. One, when the cat moved slowly, his vision was normal. So his vision could not have been physically wounded. But two, when he sped up, his visual tritinuum collapsed. He then lost access to one of his three, visual tritinuum points.

Obviously, I couldn't ask him questions. So I had to rely on what I could observe. Knowing this, I set about creating experiments to see what I could learn. What happened? As I said, when I left, the cat was fine. And at this point, all I'll say is that I let my curiosity guide me.

Do Tritinuums Actually Connect the Mind to the Body?

So far, I've explained how tritinuums create connections between us and our world. But they also create bridges between our bodies and our minds. Literally, they're what connects our physical selves to our inner, mental life. Obviously, this makes tritinuums very special.

Say you're a scientist who favors the empirical method (story-based, visible science). Say too that you want to use constellated science to explore how tomato plants grow. To do this, you'd look to create an empirical tritinuum. Empirical tritinuums are triangles of fractal continuums.

Here, the points which function as eyes are empiricism's two essential qualities. "Changes in visibility," and "Physical changes." Here the visibility eye would observe changes in what you can and cannot see about how tomato plants grow. The physical change eye would then monitor the story of how tomato plants physically change over time. Between these two eyes you'd then create a fractal continuum. This fractal continuum would be where you blend, store, and process what these two eyes observed. To create this continuum, you'd use what these two eyes share—their common mental purpose. In this case, you'd use your desire to see how tomato plants grow.

Once joined then, these two eyes—and this mental continuum—would become fractal. In effect, you would have created a third geometry. At this point, you'd be ready to actually make some observations, as without them, this continuum would exist only in your mind.

To do this, you'd need to identify a focal point for these two eyes. Obviously, this third point would be the tomato plants themselves. Here, each eye would make its own eye-to-tomato connection. And each eye-to-tomato connection would collect and feed sensations back to the mind.

Can you see how these two collection continuums would bridge your body (physical world) and your mind (mental world)? On the body side, your five senses would create a pool of physical, real world information. On the mind side, your mind would create a sensed continuum of visibly changing understanding—the story of this data collecting. At this point then, the data collecting and processing would begin. Here, you'd cycle between two tasks. The body's task would be to collect and feed physical sensations back to your mind—connecting your body to your mind. And the mind's task would then be to empirically contrast and compare what's visible with what is changing, then tell the body to collect more data. This would connect your mind to your body.

Cycling between these two tasks is what creates the feedback loop I've been referring to—the sensing<->understanding cycle. Here the cycle would have two goals. The first would be to fill the mind with all the ways tomato plants grow—the second to explore this information, looking for new patterns.

Finally, if successful, these newly-discovered, mental patterns would be put to use. For one thing, they'd be used to predict ways to improve the process of growing tomatoes. For another, they'd be used as the basis for measuring the success of this process. And so on and so forth. On and on. Physical to mental. Mental to physical.

Can Tritinuums Be the Source of Perception Itself?

Yet another thing to realize about tritinuums is how they enable us to see ranges of possibilities. Here these ranges are not ordinary ranges. They extend out to theoretical extremes. In part, this is what makes pairs of complementary opposites so important—and why a scientific method must include both the third and fourth geometries. It's the third geometry which creates this continuum of choices. And it's the fourth which allows us to see the third.

In effect, tritinuums enable us to access the *full scope* of whatever we explore. Tritinuums also give us the ability to narrow our scope. This allows us to focus in on whatever we're sensing, or know why we can't.

For instance, it's tritinuums which reveal why it's hard to find answers which satisfy opposites like science and religion (the sensed continuum from the measurably rational to the unmeasurably irrational). And the same goes for common sense opposites, like the advice to "focus on the journey, not the outcome" (the sensed continuum from being to doing).

Tritinuums also explain why the same person can't be a lover and a friend to someone (the sensed continuum from me first to you first relationships). As well as why you can never love two paintings—or two people—including children—equally, at least not in the real world (the sensed continuum from most loved to least).

More important still, they also reveal some pretty important truths. For example, despite the cliche which warns us against reinventing the wheel, in truth, tritinuums tell us we must keep reinventing things. Why? Because static truths—like all single points—do not remain visible, let alone viable. Thus in order to keep seeing these patterns, we must keep reinventing (and rediscovering) them—both mentally and physically.

Are you beginning to see how tritinuums benefit us? They alone give us the ability to discover the scientific nature of things. They also give us the ability to feel these discoveries personally, making tritinuums the heart and soul of human perception. Here, by *perception*, I mean "intelligently processing sensations."

In addition, tritinuums allow us to anchor our scientific viewpoints, connecting us—the scientists—to whatever we explore. This gives our searches focus and purpose, and makes us conscious of both ourselves and our world. Hence this geometry's title; "the consciousness creating tool."

Can Tritinuums Identify—and Heal—Wounds?

Now let's talk more about a problem I mentioned earlier. The problem? That we can sometimes become unconscious during the sensing<->understanding cycle. And yes, considering we get our ability to be conscious from this cycle, this may sound odd. Like all things in nature though, consciousness has two sides.

In effect, becoming conscious has what some might see as an evil twin. Here the evil twin is becoming unconscious. Moreover, whenever we become conscious of something, the evil twin appears. At which point we feel stuck—should we keep looking, or should we look away?

Why does this matter? As I've said, to connect to the physical world, we must first focus on a single point. But if we linger too long and don't look away—this point becomes invisible to us. Psychologists call this, "dissociating," meaning, we disconnect from the world. And for most of us, this happens on and off, in minor ways, all day long.

In part, this is just an innocuous aspect of human nature—we tend to get tired and bored in the absence of change. But this also happens whenever we relive a wound. Indeed, this is largely why it hurts so much to relive wounds. On the one hand, we want to keep looking at the terrible thing. And on the other, we want to look away.

More importantly, the more we struggle to resolve this indecision, the more blind we become. Both choices have their good side and their bad. This makes it hard to make a choice. This results in us freezing in a state of indecision, which only adds to the pain. It also causes us to relive our wounds the same way, over and over again.

Now as I've said, we'll talk at length about the nature of wounds in a future chapter. For now the thing to focus on is this state of indecision. Specifically, the idea that whenever we relive a wound, we freeze in this state of indecision. This unanswered question—and the unconsciousness it creates—then renders all other choices invisible to us.

Worse yet, since human intelligence is based on patterns not on situations, this blankness is context independent. Any time we relive the pattern, we'll relive the wound, no matter where we are. Know this is what makes wounds so difficult to find—and so hard for therapists to heal. Because wounds are patterns of experience and not the experiences themselves, there is no way to predict when and where we'll encounter them. Relive the pattern and you relive the wound.

For example, take the idea that each time we fall in love, we tend to overlook certain negative traits in the other. This happens because our own wounds blind us to the other person's faults. We simply can't consciously experience these faults—even when we see them. At least, at first. But eventually, we hurt so badly, we can't continue to excuse things. At which point, we either heal ourselves—or we leave, only to seek out the same underlying patterns in another person.

This permanent ambivalence is also what causes people to repeatedly fight the same fights. This happens whenever both people's wounds overlap. This overlap creates situations wherein both people see no other choices. This is why, when you're having one of these fights, it feels like the whole fight is scripted. And in a way, it is. No matter how hard you try to avoid it, you do and say the same things, again and again.

Fortunately, the fourth geometry offers us a way out of this stuckness. For example, if you're struggling with one of these situations, a tritinuum could show you where you keep getting stuck. Here, you'd focus your efforts on pinpointing places where you can't visualize other choices. In effect, in these moments, this tritinuum of choices will collapse to a single point. This will then leave you with no depth perception—nor any real sense of who you, and the other person, are.

Again, in a future chapter, we'll explore the many ways you can use tritinuums to pinpoint blankness. Know these cases of ambivalence-driven unconsciousness can also happen to the tritinuum-based parts of your body. By this, I mean they can happen to your ears, nostrils, hands, feet, knees, shoulders, and so on. And each time this happens, you'll feel pressured to choose a point to focus on.

At this point in the pattern, you'll begin to feel urges to keep rescanning the stuck point. Unfortunately, you'll probably be unable to do this, either because you can't bring this point into focus—or because you can't move off of this point. Either way, you'll lose access to your choices. You'll literally become unconscious in these situations. And this will then lead you to think, feel, and behave in the same ways you always do.

For example, consider the problems people experience when filling out medical-intake forms. In these situations, these folks get overwhelmed by the myriad empty spaces. This leads them to freeze in indecision and to feel like they can't go on. In effect, trying to fill out these forms renders them unconscious.

Know this sometimes results from having been suddenly overwhelmed by a fill-in-the-blanks test in school. Whatever the source, what's important to see is this wound renders them unconscious. Once unconscious, they not only become unable to see to what's on these forms. They also become blind to *everything else around them.*

Similar things happen to people who hate math—another common school-related injury. When these folks see numbers, they freeze and go blank, and this blankness prevents them from seeing where to start. They literally lose their ability to focus on what's right in front of them—as well as *everything else around them.* They then use logic-of-the-psyche to blame themselves—reasoning that they just must be bad at math.

Know the inverse can also happen. Rather than being unable to see choices, you can get stuck staring at only one thing. For instance, take the folks who stare at the needle when they're getting a blood test. Staring at this needle renders these folks unconscious to *everything except this needle.*

This is why these folks expect the needle to hurt so badly. If the only thing they feel is the needle, it's painful as hell. Likewise the folks who discover a minor dent in the door of their new car. In most cases, they can't stop looking at this dent. This then causes them to become unconscious to *everything except this dent.* And in these moments, this dent becomes the only thing which exists in their world. Painful, to say the least.

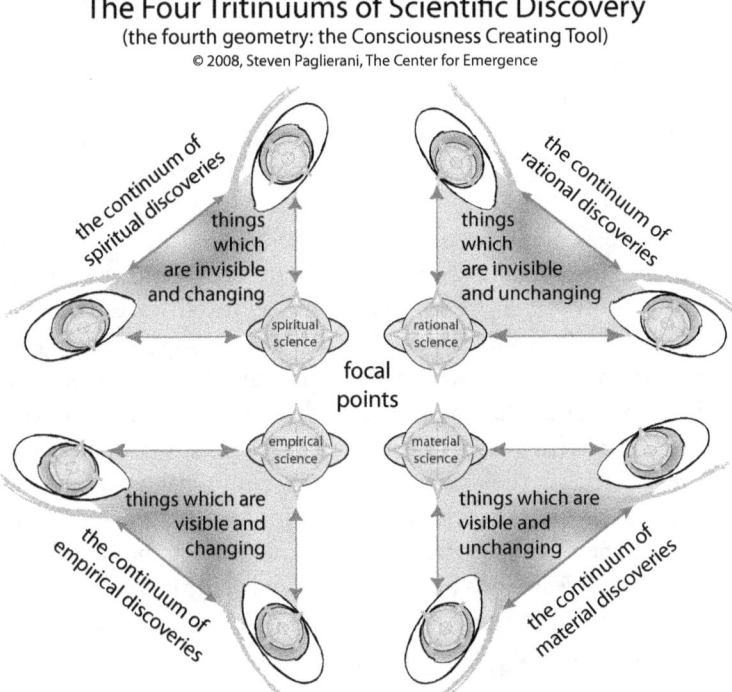

Here, each tritinuum has three interconnected single points, each of which creates six directions and three connections.

Can Tritinuums Firm Up the Soft Sciences?

Another thing a scientific method must do is enable us to firm up "soft sciences." Here, we call these sciences "soft" because they suffer from the "measurement problem." In other words, most of their measurements translate poorly to the real world. This leads to mathematical absurdities like Shrödinger's cat—a poor feline which quantum mechanics claims is simultaneously living and dead.

Realize, much of this problem stems from the present method's reliance on statistics. Using statistics enables proponents of this method to claim they have answers when they don't. Luckily, constellated science uses a scientific measurement math which, even in the real world, is 100% accurate. So much for basing science on "partly" and "chance of."

Then we have the idea that the main goal of science is to discover the nature of things, by seeing connections. This means, scientific measurements must do more than be accurate. They must also guarantee we'll discover new patterns. In addition, each of these patterns must be both theoretically sound and have real world value. Know that constellated science manages to pull off both.

For example, take the psychologically obtuse idea of disciplining children. Admittedly, this topic is controversial, leading many couples to argue as to what method is best. If you place these arguments on a tritinuum however, a pattern arises. One person is usually overinvolved—the other, uninvolved.

For example, say a wife constantly complains that her husband never disciplines the kids. Perhaps she is always shouting at the kids, then putting her husband down when he suggests she ease up. In cases like this, it's common for a husband to eventually back away and do nothing. But if you juxtapose these two viewpoints, the problem becomes clear. They're locked in a cycle of conflict, wherein neither sees the other's viewpoint.

The solution? They each must learn to see the world through the other's eyes. Mapping their differences on a tritinuum can be a good way to begin. Here, as soon as you do, this tritinuum creates a scientifically sound way to view the problem. And if you can get them to look, the solution becomes clear. She's being too harsh with the kids. And he's being too harsh with her. And they each need to be more like the other—she, less critical about his suggestions; he, more involved.

Can Tritinuums Eliminate Scientific Bias?

Of course, for solutions like this one to work, the people involved must have open minds. This brings us to the next problem tritinuums solve—scientific biases. And yes, we spoke previously about how the third geometry can open minds. What tritinuums add is the ability to address a scientist's personal closed-mindedness.

In a way, all prejudices, scientific and otherwise, are personal closed-mindedness. For example, say you're a scientist who's been asked to explore people's thoughts regarding large-sized, new homes. Unfortunately, you're biased against buying these homes—you feel they're all Mac-mansions. So you bias your hypothesis toward the idea that they all come from cookie cutter molds. What a setup for failure!

Is it possible that you're right—that there are no large-sized, well-made new homes? Logically, this seems like a stretch. Unfortunately, logic alone is rarely enough to open a mind. However, if you draw a tritinuum containing all possible reactions to large-sized new homes, you just may gain enough insight to save your study.

What would this tritinuum look like? For one thing, it would contain the full spectrum of contradictions. Reasons why people think large-sized, new homes are crap. Reasons why people think they're excellent. Indeed, seeing these contradictions on paper may just push you towards a mind-opening surprise. After all, the evidence for your bias is right there, on the page, and bias skews any science.

Another example would be scientific claims made about political integrity. For instance, some folks claim there are no honest Democrats—that they're all corrupt and lie. Here again, the absence of any complementary opinions is evidence this person's mind is closed. Where are the honest Democrats? There are none?

A political tritinuum will quickly reveal the visual contradictions here. Does your belief mean all Republicans are perfectly honest—that they never bullshit or steal? Overall the thing to see here is that scientists are—after all—only human. No surprise then that they can sometimes have strange beliefs, like that all men want just one thing; sex—or that all women don't want sex; that they're all frigid.

Obviously, these beliefs are biased, limited, and one-directional. Which is why the visual contradictions contained in a tritinuum can often provoke ahas.

For instance, consider a tritinuum which included what the world would be like if no men liked sex—or if all women were insatiable.

Clearly, tritinuums can address biases—both scientific and personal. To do this, just ask the scientist to insert all the possibilities into a tritinuum. If he can, then he's ready to scientifically explore his topic. But if he can't—or if he sees only a limited number of choices, then he's unlikely to make any significant—let alone scientific—discoveries.

Can "Good Things," Close Minds?

Did you notice how I used complementary closed-mindedness to create the tritinuums I referenced above? This is a good example of how inserting bad things into tritinuums can open minds. Of course, if you only see the bad choices, your mind will stay closed. This state of mind even has a name—pessimism.

The thing is, there are also situations wherein people see only good choices. This, too, has a name—idealism. Here, closed mindedness manifests as being closed to the bad choices. Strangely, there are books which claim this second attitude is the path to getting what we want in life—"if we think only good things, we'll only get good things."

Setting aside the idea that doing this is humanly impossible, it also makes for really bad science. Indeed, since extreme versions of bad and good perspectives can both end up closing minds, both perspectives will likely bias your beliefs—and contaminate your science.

For example, consider a tritinuum which contains beliefs about how cops treat people. This tritinuum might range from *all* cops endorse brutality to *all* cops are gentle and kind. Or consider a tritinuum which contains beliefs about how women treat people. This tritinuum might range from *all* women are critical bitches to *all* women are naturally nurturing.

In both cases, the first perspective focuses entirely on the bad, while the second focuses entirely on the good. Hopefully, it's obvious to you why these tritinuums have problems, starting with the word, *all*.

All implies the outcomes are fixed before you even start, hardly a way to conduct science. No coincidence, nineteenth century personality theorist, Pierre Janet, used this same concept to define wounds. He described wounds as "fixed ideas." Things which are all good and all bad are fixed.

Fortunately, tritinuums show us how the problems in closed-mindedness can exist in both good and bad things. Here again, we see an example of how tritinuums reveal biased views.

Can Tritinuums Reveal the Good in Negative Things?

Admittedly, Janet's idea—that wounds are "fixed ideas"—can at times seem quite brilliant. However, according to his definition, wanting to "always be healthy" is also a "fixed idea."

Certainly, it's good to *want* to be healthy. Then again, *always* wanting to be healthy sounds a bit off. Can always being healthy really be a good thing? It makes me wonder, is this why people with excessively positive outlooks beat themselves up when they get a cold? Moreover, can this kind of self-induced stress actually make things worse?

Clearly, science tells us stress is a major factor in getting ill. And thinking you should always be healthy will definitely generate stress. This time, the word *always* points to the problem. Like the word *all*, the word *always* refers to a single point.

Is your mind closed then when you make these kinds of proclamations? Good question. In a way, it is. But in this case, there's something that makes this the exception to the rule. In this case, it's more like what it's like to be certain I'm seeing an acorn, or a potato, or a cloud. What I'm certain of is the recognizable part of the fractal pattern I'm seeing.

Confused? The key here is to remember how I defined fractals.

Fractals are *recognizable patterns that **always** repeat differently.*

This means, there is one kind of certainty that is scientifically desirable. So while it may sound to some like I have been contradicting myself, in truth, I've just been engaging in the exception that proves the rule. The recognizable part of fractals is the one thing science can be certain of.

As for health, since health is not fractal, any certainty involving health will fall outside the certainty exception zone. Here this belief simply ends up being yet another a single point—in this case, an impossible real world goal.

Fortunately, the tritinuum of health and illness reveals this problem. Like the word *always*, both health and sickness must have an upside and a down. By this, I mean, there must be a good side to getting sick—and a bad side to staying well. And to see this in action, consider what happens to you when you get a cold.

Obviously, no one likes getting a cold. You get tired, sore, cranky, and you sniffle. The thing is, scientists tell us that getting colds revs up our immune systems. Some scientists even suggest that without getting colds, we humans might never have survived—that without these minor tune-ups, our immune systems would eventually atrophy.

What keeps us from seeing these kinds of important, personal truths? No one has previously seen logical geometry—so we haven't learned to ask the right questions. And in the case of tritinuums, we learn to ask about the good and bad sides of all things. And this leads us to discover strange things like that we need to get sick to be well.

Using the Fourth Geometry to Clarify Goals

Another way tritinuums benefit us is that they help us to understand goals. For instance, many people who value physical fitness focus entirely on getting fit. Admittedly, at times, this focus can be a good thing—being goal oriented can lead to success.

But if we focus too much on idealized states, like being fit, we miss the beauty in—and the value of—the days when we don't go to the gym. Those slow, calm vacation site-seeing trips and those leisurely walks in the park. Who wants a life wherein all you do is work!

Conversely, some folks focus only on the journey toward health and fitness, all the while, never acknowledging their successes. On the positive side, doing this gives us an appreciation for the struggles in everyday life. Overdone though—such as when we spend more time measuring our fitness than enjoying life—life becomes one unending, ever-widening evaluation.

The point is simple. To experience choices, we must see three points. Two form a fractal continuum containing our choices—and one forms our vantage point. By inference then, we could define scientific viewpoints as states wherein you can see an unlimited number of choices—both positive *and* negative. Moreover, knowing that tritinuums give us the ability to see these choices gives a good way to test for whether a tritinuum's been properly done. The choices on properly done tritinuums always include both the presence and absence of both good and bad choices.

Admittedly, most times, following this dictum is easier said than done. If you understand the nature of tritinuums however, you understand why seeking purely positive choices is an unscientific view. Trying to be the perfect parent? Aspiring to an entirely pain free life? Tritinuums teach us that the inability to see the good in negative choices is just as unscientific as the inability to make positive choices. This makes tritinuums an essential ingredient in any scientific method. And in life.

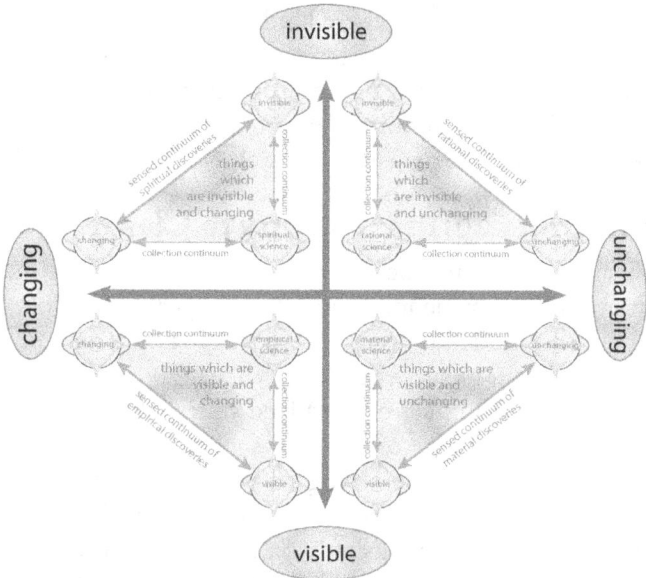

The Crossed Continuums of Constellated Science
(the fifth geometry: the Expanding Tool)
© 2008, Steven Paglierani, The Center for Emergence

The Fifth Logical Geometry: Crossed Continuums

Now we come to the pivotal point in this sequence of six forms. I say this as it's the fifth geometry which sets the discovery process in motion. What I mean is, in the previous four geometries, our main job was to gather observations. Whereas in the fifth geometry, we begin to arrange our observations into scientifically, significant patterns.

What makes a pattern, "scientifically significant?" One thing. It reveals fractal patterns while omitting none of what's been observed. In effect, every single piece of evidence must fit correctly into this pattern. No adjustments allowed. No ad hoc rules. No leaving evidence out.

Strangely, the present scientific method sees nothing wrong with omitting things. If data doesn't fit a hypothesis, oh well. Just call it an anomaly and leave it out.

For instance, take medical science. Medical scientists crucify folks who make claims which aren't evidence-based. And in many cases, their vigilance saves lives. At the same time, when medical researchers observe things which don't fit into their theories, they often see nothing wrong with omitting this evidence.

For example? The best one I know of involves what medical researchers call, *placebos*. Placebos are supposed to be things which have no genuine medicinal value. The odd thing is, sometimes placebos work—sometimes "sugar pills" heal people's illnesses. Oddly, I've not read of a single scientist whose investigations into this were taken seriously.

Aren't placebos *evidence* then? Apparently they're not. Nor are cases which doctors refer to as "medical miracles." So what are legitimate examples of unexplained healings or remissions. Are they really *miracles*? Really? Who's being unscientific here?

Admittedly, ethical scientists at least acknowledge such omissions— perhaps with a footnote, or a mention in their opening statement. The more honest may even apologize for their inability to account for these unexplained events. But the less ethical will do just about anything they can to get you to believe these omissions don't matter—anything from hiding behind their titles or popularity, to attacking another scientist's character.

Then there are those who believe spiritual beliefs have no part in science. I wonder what they think of Einstein's crediting *intuition* or Niels Bohr putting the Tao Symbol in his family crest? Perhaps these two Nobel Prize winning physicists were pseudoscientists? If so, then why are they considered two of the greatest physicists of the last century?

Then we have the folks who feel they must protect us from "radical" ideas. Somehow these folks never notice, most scientific breakthroughs once fit this category. Most of these things were also once called *pseudoscience*. But we suddenly call them breakthroughs when these things turn out to be true? What exactly have we broken through? Closed-mindedness?

Then we have two of the more egregious examples. The first happens when scientists ignore the difference between direct observations and statistical "evidence." The second happens when scientists take logic to be evidence. Statistics repeatedly and reliably fails to predict individual cases. And logic alone offers close to no inherent, predictive power.

In both cases, scientists substitute guesses and "almosts" for real evidence. That doing this doesn't trouble more scientists never ceases to amaze me. And yes, I believe things like statistics and logic have a place in legitimate science. Statistics is a really good way to find starting points. And logic is great for seeing through false conclusions. But as we've seen, mistaking starting points for final results only leads to problems. As does claiming a starting point is all we need to explore.

What happens when you try to ask about these omissions? The first thing people usually do is point out where we'd be if it weren't for science. And to be honest, they're right—without science, we'd be lost. I sincerely mean that. But if your method permits you to omit evidence in order to get results, then I believe you forfeit your right to call your work *science*.

I realize, to the more conservative scientists, these claims can sound harsh. Is the primary goal of science actually to find patterns of connections in *natural* things? And can natural things only be considered natural when taken whole and in situ? According to constellated science, for a method to be considered scientific, it must account for more than just cherry-picked data and wished-for ideas. This method must make sufficient observations—and account for every single piece of evidence observed.

Admittedly, doing this with the current scientific method would make a slow process even slower. But since legitimate science looks to discover the nature of natural things, if you omit parts of this nature, can you call your work *science*? I think not. Well-intentioned? Maybe. But science? No way. So if I'm guilty of something here, it's simply that I think it's time to raise the bar. Practicing sixteenth century science in the twenty-first century just sounds wrong.

Constellating These Observations into Scientific Patterns

As for how the fifth geometry addresses these problems, know this is what we'll be looking at next. Keep in mind—while we explore this process—that the heart and soul of constellated science is arranging observations into scientific patterns.

Ironically, for all its power, to me, the fifth geometry seems simpler than the fourth. After all, as we've seen, its just two crossed, fractal continuums. At the same time, there are four connections between its four end points and the intersection of these continuums. And these four connections do make this geometry a four-point set with four directions.

In truth then, despite its simple appearance, the fifth geometry is truly complex. In it lie the seeds of four *complementary* tritinuums. Imagine? Four completely, complementary tritinuums. Admittedly, in the fifth geometry, these tritinuums are only implied. They come into being only in the sixth and final geometry. As does the point where these four tritinuums intersect—the all-possible-viewpoints point—the heart and soul of the entire search.

Finally, as we'll see in the next chapter, there's good news on the horizon. Once you create a fifth geometry, the sixth geometry comes into being largely on its own. Indeed, once you finish a fifth geometry, in many ways, the sixth emerges almost magically. Here, I'm referring to artificial intelligence I spoke of earlier—an intelligence mainly rooted in the unbroken sequence of logic which threads through all six geometries.

How exactly does this work? You're about to find out.

The Six Logical Geometries
(the Connection-Discovering Tools)
© 2008, Steven Paglierani, The Center for Emergence

a single point the genesis tool

a linear continuum 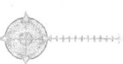 the measuring tool

a fractal continuum 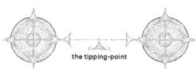 the possibility defining tool

a tritinuum 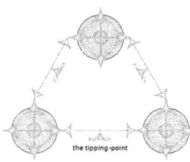 the consciousness creating tool

a crossed continuum 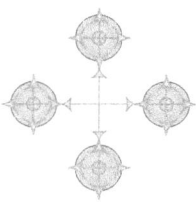 the expanding tool

a quadtinuum 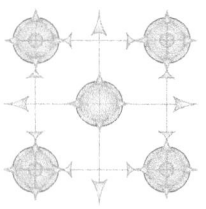 the discovery tool

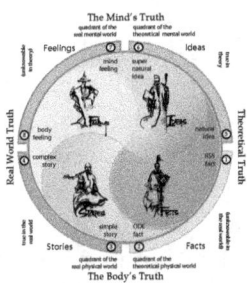

Section 1 - Chapter 9

Using Logical Geometry
(mapping the natural world)

What are the Four Map-Making, Action Steps?

Now let's put what we've been learning into action, by making a few maps. We'll map two theoretical, natural sciences—the four Geographic Directions, and the four Arithmetic Operations. And we'll map one practical, natural science—a Guide to Career Advancement. In all three cases, we'll use constellated science's four map-making, action steps. These four steps are:

Create the Fifth Geometry: the First Action Step

[1] Draw an equal-arm cross, define two axis-defining questions, and assign these questions to the two axes of the cross.

Create the Sixth Geometry: Action Steps Two, Three, and Four

[2] Answer the two questions. Then assign the answers to the cross.

[3] Define the four kinds of discoveries (by combining adjacent answers) Then place their names in the map's, "classes of discoveries" quadrants.

[4] Define the four natural states (by combining adjacent discoveries). Then place their names on the map's, "natural state" boundaries.

The thing to pay close attention to here is how I've divided this list. The first part defines a map's fifth geometry; the second, the sixth.

Know we'll use these same four steps each time we create a map. Here, the easiest part will be drawing the cross. Then the hardest part will come next; discovering the map's two tipping-point based questions.

Realize, sometimes as I begin I think I can already guess a map's four kinds of discoveries. Or sometimes I'm pretty sure I can guess a map's four natural states. The thing is, even if I'm right, I still need to find the two questions which lead to these discoveries or states. And if I skip these questions? Then it means I've based my map on unfounded assumptions. In doing so, I forfeit my right to call my work *constellated science*. Why? Because in constellated science, nothing is ever ad hoc—inserted just to make things work.

Mapping the Four Directions: Action Step One

Of course, sometimes you get a break and these questions come to you straight away. Take the map of the four directions. I knew north, south, east, and west had to be in there somewhere. In the end, I came to realize these four directions could exist only in theory. This meant my map had to have a real world reference point—a "point from which I could stand and look at the rainbow."

Ten minutes later, I had this map's two tipping-point based questions. All I had to do was to orient my map to the Northern hemisphere sky at night.

[Question 1] Can I see (am I facing) Polaris—the North Star?
[Question 2] While facing Polaris, is it moving to my left or right?

Next, I needed to assign each question to an axis. To do this, I looked for parallels between these questions and those in the prototype map. Here the *visibility* question always gets assigned to the vertical axis, while the *change* question always gets assigned to the horizontal axis. Obviously, the "can I see" question refers to visibility. And the "is it moving" question refers to change. Thus the "can I see" question gets assigned to the vertical axis. And the "is it moving" question gets assigned to the horizontal axis.

Mapping the Four Directions: Action Step Two

Next I needed to begin to create this map's sixth geometry. To do this, I needed to answer this map's two questions—then place these answers at the ends of each cross arm. The answers to the two questions?

[Answers-Q1] "I am facing (I can see) Polaris," and, "I am not facing Polaris."

[Answers-Q2] "As I face Polaris, I see it moving to my left," and, "As I face Polaris, I see it moving to my right."

At this point, I needed to parallel this map to the prototype map. Why? Because as I've said, things which are *naturally visible* (things you can see directly) always get assigned to the bottom of the vertical axis. And things which are *naturally invisible* (things you cannot see directly) always get assigned to the top of the vertical axis. Then things which, *by nature, refer to change* always get assigned to the left end of the horizontal axis. And things which, *by nature, do not change* always get assigned to the right end of the horizontal axis.

As I considered this, I realized I needed to assign—*I am facing Polaris (I can see it)*—to the lower, vertical axis. I then assigned—*I am facing away from Polaris (I can't see it)*—to the upper, vertical axis. Then I assigned—*while facing Polaris, it is moving to my left*—to the horizontal axis, right-end—*change*. And finally, I assigned—*while facing Polaris, it is moving to my right*—to the horizontal axis, right-left—*no change*.

Mapping the Four Directions: Action Steps Three & Four

In step three, I then combined adjacent answers in order to define the four kinds of things we can discover about direction—the four classes of discoveries. Unfortunately, when I tried to do this, I hit yet another wall. For one thing, I had no idea what to call these things. So at first, I tried using *east*, *west*, *north*, and *south*. But when I tentatively placed them into the four quadrants, it looked nothing like cartographer's maps. In fact, it looked all wrong.

I also had no clue what to call this map's four natural states. So at this point, I decided to try using east, west, north, and south. Then, as I tentatively drew this, it all began to make sense. Together these states create a natural boundary within which all directional knowledge lies.

This idea—that each map's natural states must define the limits of where we should search—is one of the significant improvements the new scientific method makes over the old. Can you see why? For a scientific search to be scientific, it must have clear and certain limits for where to look. At the same time, it must allow for all possible observations within these limits. Strangely, the current scientific method accomplishes neither. Nor does it mention this flaw.

Okay. Now I knew that east, west, north, and south were the names of the four natural states. But I still didn't know how to orient these four words. Should I just parallel normal maps—putting north at the top? I knew if I based this placement on an assumption, it would invalidate the entire map. I then decided to do what all good scientists do. I did an actual experiment. I stood behind my home and faced Polaris. Doing this gave me north. This also gave me south. South had to be behind me. I then looked to my right and realized that this was east. Then as I looked to my left, I realized it all made sense. To my left, was the west.

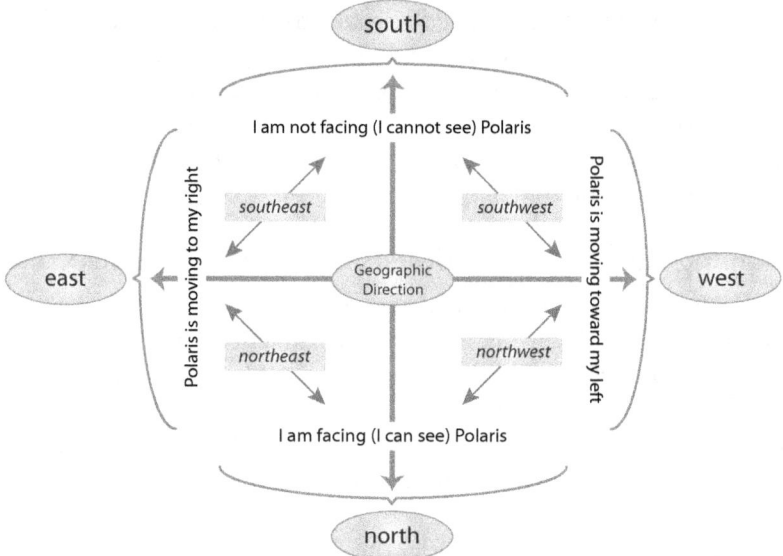

Okay. I had real world observations to guide me. But I still needed to reconcile these observations with the prototype map. And as I pondered how to do this, I had another aha. If I let my experiment guide me, the visible point on the cross had to be north. So north had to be at the bottom of the cross. Moreover, if I was facing north, east had be to my right, west to my left, and south had to be behind me. This meant south was invisible to me. Aha! And in my experiment, it was.

I now labeled the cross's end points. But I still hadn't named the four classes of discoveries. Then it dawned on me, I might be able to reverse engineer these names. To do this, I'd need to split the difference between each adjacent pair of natural states. And as soon as I did this, I had the four names—northeast, northwest, southeast, and southwest.

What did I learn from making this map? Two things immediately come to mind. One—that the "rainbow principle" applies to the placement of all things in maps. Two—that there will always be two ways to arrive at a map's classes of discoveries. With point one, I saw that to place things in maps, you must always place them *relative to something in the real world*. And with point two, I realized, there are two ways to come to answers. We can either combine adjacent answers (the normal path)—or we can split adjacent natural states (the reverse engineered path).

Even with all this though, at this point, I still worried I'd somehow missed something. So I kept going over these placements in my mind, rechecking my parallels and logic. That's when I saw it—the unbroken chain of logical geometry in my map—questions to states. This clearly proved that east, west, north, and south had to be states.

What did this mean? It meant these four natural states must be the theoretical prototypes from which all directions derive. In effect—east, west, north, and south represent the limits of what we can explore in geographic maps. How do I explain this map's odd-looking result where north is at the bottom? It's simple. According to the rainbow principle, all things in the real world constantly move. Thus ideal states like north and south must always be moving as well. This made the four prototype directions literally resemble the ends of a rainbow. And like being unable to reach the ends of a rainbow (because they exist only in relation to us), it's literally impossible to determine an ideal direction. So the orientation had to come from picking a real world point.

Realize—prior to this map—I had no idea maps and rainbows shared anything at all—let alone that for a map to be real, it must orient itself relative to something in the real world. In effect, to where you're standing.

Are you beginning to see the potential for discovery here?

Must All Maps Physically Parallel Each Other?

Now let's talk about why all of constellated science's maps must parallel each other. This includes even maps from seemingly unrelated sciences, such as psychology and physics. No coincidence Freud based his whole work on the idea that the mind must be governed by the same laws as physics. He saw both psychology and physics as natural things, and believed they should both derive from the same underlying patterns.

Constellated science is also based on this same belief. And by now, the proof for this belief should be obvious—all of constellated science's maps employ the same set of parallels. Indeed, by paralleling all aspects of all maps, you guarantee all maps will be logically and geometrically consistent. Moreover, if you allow no breaks in the sequence of logical geometry, this will hold true even in cases wherein you've yet to fully grasp some of these parallels. It will also guarantee that each and every scientific discipline's work would need to agree with the work of all others. This would force all scientists to respect and help each other.

Can you imagine what that would be like?

One last thing. Are you still uncomfortable with the way I've oriented north at the bottom of this map? Why can't I place north at the top, in the same place as the cartographer's rosette? The answer? Here again, for constellated science to be a better method, it must be internally and externally consistent. This begins with making every aspect of every map parallel the content *and placements* of every other map, all the way back to the prototype map.

In my map, I can see north. So south must be invisible.

At the same time, this placing north at the bottom of the map changes nothing with regard to scientific accuracy. It truth, by complying with the rainbow principle, it actually adds to this map's accuracy. The point is, adhering to this paralleling requirement means your discoveries will be comprehensive, integrated, and sound. Moreover, while at this point my claims may still seem exaggerated, I cannot emphasize this enough.

Construct a map properly and none of your discoveries will ever contradict your other discoveries. Ever. Thus you'll never need to omit or overlook any data in order to get a result.

Think about that.

Can the Four Arithmetic Operations Be Mapped?

Now let's look at the second theoretical example—the map of the Four Arithmetic Operations. Here, I'm referring to the four ways we can count things—addition, subtraction, multiplication, and division. Know that this time, one of the map's two tipping-point based questions came to me right away. This question? "Does this operation result in an increase?" Here addition and multiplication tip *yes*—while subtraction and division tip *no*.

Can you see how this question tips the four arithmetic operations into two, clearly defined groups? Addition and multiplication go together. Subtraction and division do as well. All maps stem from a similar pair of questions, and the first question always divides the four things into two pairs. Then the second question must divide each of these pairs so you end up with a second set of two pairs.

Confused? Don't worry. It will all become clear soon. And my point is, I needed the second question. Unfortunately, no matter how hard I tried to find it, I had no clue what it could be. For a time, I repeatedly reviewed what I knew. I knew there were exactly four operations. And I knew these operations all counted things. I also knew adding and multiplying resulted in an increase, and subtracting and dividing, in a decrease.

Intuitively, I sensed that in the second pairing, addition went with subtraction—and multiplication with division. Adding and subtracting are both easier to learn—learning to multiply and divide are much harder. At one point, I even suspected this question might have something to do with the rate of change each operation caused. Multiplying and dividing often change things more quickly than adding or subtracting. But over time, I realized this was just a dead end, and I grew more and more frustrated.

Were these operations, answers? Classes of discoveries? Natural states? Until I had both questions, I knew I couldn't know. Two years passed. Then finally one morning—in the shower—I realized what I'd been overlooking. In hindsight, I've often asked myself if the water made me more conscious that day. Or perhaps being half-asleep allowed me to circumvent my logical mind? Whatever the case, as I considered the problem, I realized that the first question I already had referred to change. And since no map can have two *change* questions, I realized I needed to focus on the axis for which I had no question—the vertical axis; *visibility*.

As this came to me, I face-palmed myself. For two years, I'd been forgetting to parallel the prototype map. Here, I'd had the first parallel all along—the *change* question—increase or decrease. But I'd failed to seek a parallel to the prototype map's other question—the *visibility* question.

Of course, knowing what I needed didn't mean I automatically had the second question. In truth, finding this question still took me weeks. In truth, I had no idea how to connect math to visibility. In the end though, knowing what to look for gave me enough to complete my search.

What was the second question? Duh! "Is this *entire* operation visible?" In other words, can you *see* this operation from beginning to end? To my delight, I realized this question clearly divided the first two groups into a second pair of groups. With addition and subtraction, you can see what's happening from start to finish. With multiplication and division, you can't—part of the process is invisible.

Admittedly, thinking about arithmetic operations in this way is beyond counterintuitive. Indeed, if your head is hurting right now, don't be hard on yourself. Mine certainly has. But if you want to see what I mean by "completely visible," consider the following story.

Say there's a table in front of you and that on it are two groups of five apples. Now ask yourself this. What would it mean if I asked you to add these two groups together? Hopefully, the answer is obvious. Adding them together would mean shoving the two groups of apples together, then focusing on the result—a single group of ten apples.

Now consider what subtracting five apples from this group of ten would mean. In this case, you'd look at the group of ten apples, then move five away. Then you'd focus on the result—the remaining group of five.

The point? In both cases, you'd clearly be able to witness the entire process. Nothing magical going on here. Everything that happens is visible.

Contrast this with multiplication and division, where this would not be the case. For example, say I asked you to multiply one group of five apples by the number two. Even though you'd still end up with one group of ten apples, even saying this is odd. More to the point, unlike the process of addition—where you'd see the entire process—this time, part of what happened would occur *only in your head*. Why? Because, while you'd see the five apples, the "two" does not refer to apples. Nor to anything in the real world, for that matter. This "two" is just an abstract. And you cannot picture abstracts.

Is this still not making sense? Then consider this. As we just saw, addition and multiplication can both result in identical outcomes. Moreover, in this case, both results are arithmetically correct. However, although these two outcomes are both the same, we can't just say that multiplication means moving one group of five apples over to another group of five. Something other than "moving" is happening here. Something invisible. Something abstract.

This then was my big realization—that neither multiplication nor division can be visualized from beginning to end. In truth, something happens during both these processes which our eyes cannot see. Something almost magical.

What does this tell us about arithmetic?

For one thing, it explains why children must memorize the times tables. Multiplication involves an abstract step. And as I said, we cannot visualize abstracts. Thus the only way to learn to multiply is to blindly memorize the hundred root outcomes. In one sense, for us to learn to multiply, we must take a teachers word for this part. And in another, learning arithmetic is clearly biased toward kids who can manage abstracts. Kids who are more practical—well, they suffer.

This incomplete visibility also explains why multiplication and division are harder to learn than addition or subtraction. And while I'm sure you knew this all along, I'm willing to bet you had no idea why. Amazingly, it all comes back to that it's harder to do things we cannot visualize.

So what does this map look like? And what are the four arithmetic operations—answers, classes of discovery, or states of truth? More important, what else does this map have to teach us about arithmetic? To see, we'll need to use the four action steps to create this map.

We'll begin by drawing the cross.

Mapping Arithmetic: Action Step One

This time, why don't you try constructing the map along with me. First draw the cross. Now write down this map's two questions.

One— is this entire operation visible?

Two— does this operation result in an increase or a decrease?

Now we need to assign these two questions to the two axes in the map. As I've said, to do this, we'll need to parallel the prototype map. Here the visibility question—*is this entire operation visible?*—will get assigned to the vertical axis. And the change question—*does this operation result in an increase?*—will get assigned to the horizontal axis.

Mapping Arithmetic: Action Step Two

At this point, we've completed action step one. We've defined this map's fifth geometry. From here on out, with each step we take, we should discover new connections.

What can we possibly discover about ordinary arithmetic?

Prepare to be surprised.

Okay. First we need to take action step two. We need to answer our two questions—then place these answers onto our map. Again, we'll need to use the prototype map to orient our placements.

In action step one, we determined this map's visibility question—*is this entire operation visible?* So to parallel the prototype map's vertical axis answers, we'll place the *entirely visible* answer at the bottom of the axis, where the prototype map says, "visible." Then we'll place the *less than entirely visible* answer at the top, where the prototype map says, "invisible." "Less than entirely visible" means part of this process is invisible.

Next we'll need to assign answers to the horizontal axis. This time, our question will be—*does this operation result in an increase or decrease?* Thus this map's two change answers will be, *increase* or *decrease*.

Unfortunately, finding a parallel to these answers is not so easy. In the prototype map, *change* goes on the left end and *unchange* on the right. So how the heck do *increase* and *decrease* parallel *change* and *unchange*?

What are Psychophysical Parallels?

Whenever I get stuck, one of the tools I use is I look for a psychophysical parallel. Here, by "psychophysical," I mean the idea the laws which govern the physical world parallel those which govern the mind. This time, we'll start with the mental concept—"results in a decrease." Then we'll look for some aspect of the physical world where decreases are normal. Amazingly, there is such a property. It's called, entropy.

What is "entropy?" It's the tendency in all things to break down over time. Here entropy is the process wherein things tend to diffuse, get disordered, degrade, lose energy, and decay. Obviously, this happens to everything in our world—from natural resources and remaining life spans to the newness of our cars and the freshness of our food. We can, of course, make interventions to keep things looking new. These things may even seem better than ever if we try hard enough. But repairing or refurbishing things is not the same as them staying new. Thus all things grow old, deform, and break down. Even us.

At the same time, the third geometry tells us that to know something, we must contrast this thing to its complementary opposite. And if logical geometry is real, this complementarity must exist for entropy as well. Admittedly, with the exception of nineteenth century genius, Ludwig Boltzmann, most scientists vehemently deny an opposite to entropy exists. At this point though, let's assume it does and see where this takes us.

To begin with, since scientists are so certain it doesn't exist, this opposite has no name. This means we'll need to name it.

For now, let's just call it "counter-entropy."

What would counter-entropy be? It would be the kind of change which transformed decay, waste, and disorder into new life, fuel, and order. No surprise, this exists. Think dead plants into coal, oil, and natural gas—worn-down mountains into new mountain ranges—and collapsing nebulae into new stars. Counter-entropy, then, is an actual natural process wherein waste and worn out things become something new. In essence, it's the real world's way of recycling everything which wears out.

It's also an expression of the natural law which says that nothing ever gets created nor destroyed. It just changes form.

If this is true though, then why do scientists deny this half of the cycle exists? Mainly because most counter-entropy happens over millions of years—and most entropy happens at a rate we humans can see. This means, despite the prototype map designating the horizontal right end as unchange, technically, this change does happen. But it only happens over enormously long periods of time—far beyond our lifetimes.

The Example Map with Labels
(The 2 Questions, 4 Answers, 4 Classes of Discoveries, 4 Natural States)
(© 2010 Steven Paglierani The Center for Emergence)

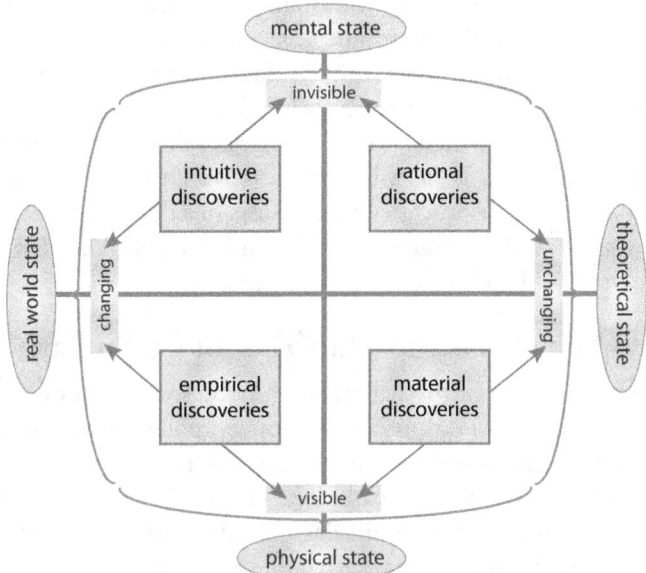

The Two Map-Defining Questions (the 2 Qualities of Existence)
(Vertical: visibility? Horizontal: change?)

The Four Placement Answers
(visible, invisible, changing, unchanging)

The Four Classes of Scientific Discoveries
intuitive discoveries (the science of invisibly changing things),
material discoveries (the science of visibly unchanging things)
empirical discoveries (the science of visibly changing things
rational discoveries (the science of invisibly unchanging things)

The Four Natural-Boundary States
(mental world, physical world, real world, theoretical world)

What Would a Map of All These Steps Look Like?

What does all this physical science have to do with a map of arithmetic? To see, consider the prototype map. Here, real world changes—those which, by nature, occur in a time-frame we can see—go to the left end of this axis. Theoretical changes—those which by nature do not change in a time-frame we can see—then go to the right.

This gives us the psychophysical parallel we've been seeking. In our map of arithmetic operations, *entropy* roughly equates to the real world process wherein something becomes less. And *counter-entropy* roughly equates to the theoretical version of something becoming more. This means, *results in a decrease* must go on the left end of this axis—and *results in an increase* must go on the right. And finally, we're ready to take action step three, where we'll need to combine adjacent answers.

We'll begin with the lower right quadrant. As we do, we make our first discovery. We discover what the four operations are—they are classes of discoveries; qualities which allow us to classify the ways we count natural things. Here *addition* is "a visible process which results in an increase." *Subtraction* is "a visible process which results in a decrease." *Multiplication* is "an invisible process which results in an increase." And *division* is "an invisible process which results in a decrease."

What Has the Map Taught Us So Far?

So what have we learned about arithmetic so far? For one thing, that the four arithmetic operations are not conceptually equivalent to the four directions. Here the four directions are natural-boundary states—the four states which define where we can make discoveries about direction. And the four arithmetic operations are classes of discoveries—the four kinds of patterns we can discover about math while exploring things in these states.

Why mention this difference? Because where we place things explains a lot. For instance, in the case of math, it explains why children learn these four operations in the order they do.

To wit, children begin to learn to count by learning how to add. Seeing the process by which things *visibly increase* is the easiest process to follow. Why? For one thing, because you can see it from start to finish. Next they learn another process you can see from start to finish. They learn to subtract. Why not learn this first? Because while you can see this whole process, anything which employs negation in its nature is harder to visualize than its positive counterpart. Where'd the subtracted part go?

Need an example? Picture getting a second cell phone vs not knowing where your cell phone is. Obviously, it's easier to picture the former than the latter. Or picture a donut vs picturing the donut hole without the donut. You can easily picture the former. But picturing the latter is impossible.

The hole is not a donut.

The point is, it's easier to learn things you can see than to learn things you can't. With subtraction, something's there. Then it disappears. With multiplication and division, parts of these processes can't be pictured at all. And yes, with multiplication and division, we can see where we start and where we end. But the part in between exists only as an abstract. In other words, we can reason what happens. But we can't picture it.

Now add all these ideas together and you will see why children learn arithmetic in the order they do. Addition is the easiest to learn. It's completely visible and includes no negation. Subtraction is the next easiest. It's completely visible, but includes negation. Multiplication comes next because it's partially visible, but includes no negation. And division is hardest of all because it's both partially visible and includes negation.

The Map of the Arithmetic Operations
(© 2012 Steven Paglierani   The Center for Emergence)

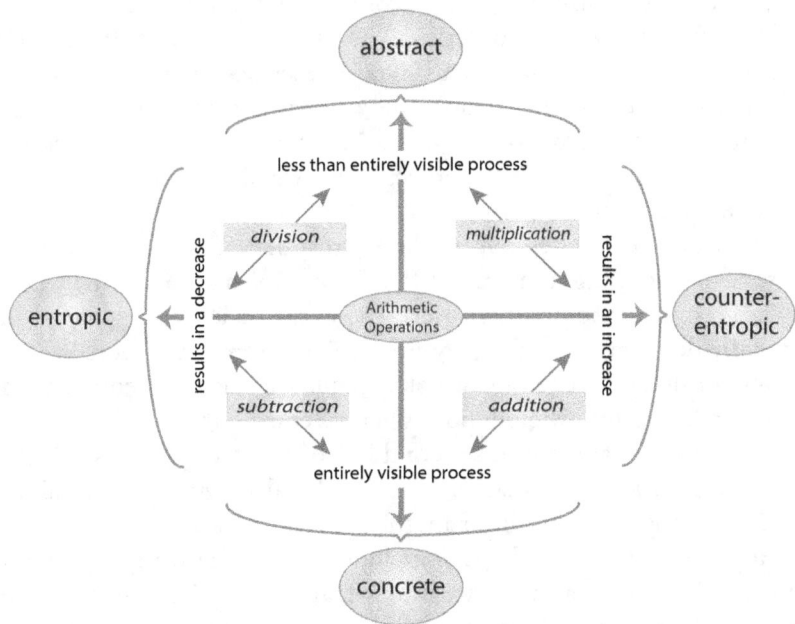

Vertical Axis Question:
Is the arithmetic operation I'm observing entirely visible?

Horizontal Axis Question:
Does this arithmetic operation result in an increase?

Mapping Arithmetic: Action Step Four

Are you beginning to see why I'm claiming that basing a scientific method on logical geometry is better? Unlike the present method, which often treats guesses as facts, this one functions more like a mathematical proof. Imagine if all branches of math had to prove every other branch's problems—both in theory and in the real world. In effect, either discrepancies would surface—in which case you'd know something is missing. Or things would elegantly fall into place, in which case it would all fit together and apply to all things.

Admittedly, I've a lot more to explain, not the least of which is to formally state this method. For now though, let's take the fourth action step so we can complete this map.

Here again, to know where to place things, we'll need to psychophysically parallel the prototype map. And when we do, we arrive at the four states in which we can make arithmetic discoveries—the entropic state, the counter-entropic state, the concrete state, and the abstract state.

Here the *entropic* state describes operations wherein stuff decreases (breaks down). The *counter-entropic* state describes operations wherein stuff increases (gets built back up). The *concrete* state describes operations wherein the entire process is visible. And the *abstract* state describes operations wherein parts of the process are invisible. In other words, they can be reasoned but not pictured.

What do these states tell us about arithmetic that we didn't know before? For one thing, they reveal the theoretical marriage which exists between physics and mathematics. These two sciences regularly and frequently demean each other, arguing which discipline is more sublime. In this map, though, we see an unbroken sequence of logical geometry—an intimate relationship wherein both sciences contribute equally.

For instance, take physic's second law of thermodynamics—that in the physical world, it's normal for all stuff to break down. In effect, this law says that as long as you don't do anything to counteract natural processes, all things decrease or divide. This includes everything from the amount of tread left on tires and the wear and tear on your jeans to the water levels in fish tanks and the amount of carbon 14 left in biological matter.

If we now look at the map of arithmetic operations, we find it contains this very idea—that when it comes to counting real world things, it's normal for whatever we're counting to decrease or divide. In order for this map to be considered valid however, it must contain the complementary idea as well. In this case, it's the idea that in theory, things can increase or multiply.

Moreover, according to constellated science, we must also find parallels to this idea in the laws of physics. And we can. Worn out tires and jeans and missing fish tank water all go somewhere. And eventually, they become something new. What makes this different from entropy then is that, for the most part, we don't live to see it. And human beings, including most scientists, are notorious for believing—seeing is believing. If you can't see it, it isn't real.

Of course, the ultimate example of this flaw in our nature is how we fear the Earth will break down. And in truth, it will. This process is happening, each and every day. Indeed, this breakdown is yet another example of entropy. But while at some point, we humans may cease to exist on this big blue hunk of rock, the Earth will likely heal itself and become new again. Hence my using the term, counter-entropy.

Are the natural states in this map beginning to make sense? Because the left side of the prototype map represents real world change (the kind we can see), I've placed entropy—nature's version of subtraction and division—on the left. Conversely, because the right side of the prototype map represents theoretical change (the kind we can imagine but not see), I've placed counter-entropy—nature's version of addition and multiplication—on the right side. Here entropy refers to any and all physical and mathematical decreases. Whereas counter-entropy refers to any and all physical and mathematical increases.

Admittedly some may find these parallels divisive—or a stretch—or no big deal. Why go to such great lengths to find these kinds of parallels? We do it because in constellated science, all laws must be universal. In effect, to be true, they reoccur throughout all aspects of life.

This is why each and every single map must parallel every other—regardless of its topic. For example, take the first theoretical map I showed you—the map of being and doing. For the map of arithmetic operations to be considered valid, it must parallel this map too. And it does. Here *being* is an entropic process—while *doing* is counter-entropic. And being is the real world norm, while doing is possible only in theory.

Know that similar, real world to theoretical world parallels exist between the map of arithmetic operations and the map of geographic direction as well. Admittedly, these parallels can be even harder to find. For instance, the planet Earth naturally rotates west to east. But the planet Venus rotates east to west. So while the real world rotation of the Earth is west to east, in theory, it may have been east to west at some time in the past.

Finally, know that this parallel between the left (real world) and right (theoretical) sides of maps can often be a good one to check for.

It can also be a downright bear to grasp.

More on this in a bit.

Mapping Arithmetic: Concrete vs Abstract States

What about the other two arithmetic natural states—the states I'm calling *concrete* and *abstract*? As I've said, the *concrete* state refers to arithmetic changes which you can see from beginning to end. Whereas the *abstract* state refers to arithmetic changes wherein you can't see parts of the process. Here, addition and subtraction are concrete operations—whereas multiplication and division are abstract.

How did I know where to position these two states? Again, I looked for parallels in the prototype map. And in the prototype map, the state of *physical* discoveries is placed at the bottom of the vertical axis, while the state of *mental* discoveries is placed at the top. So since physical truth is visible—and since mental truth is not, I placed the state of concrete arithmetic truth at the bottom of the vertical axis, and the state of abstract truth at the top.

What about the requisite parallels between this and all other maps? For instance, what about the parallels between this map and the map of being and doing? Here even a quick glance reveals the parallels. The abstract state equates to the state of mind, whereas the concrete state equates to the state of body.

On the other hand, finding these parallels in a map like geographic direction can be really difficult. How in the hell is south an abstract state, and north a concrete state?

This time, rather than give you the answers, I'd rather you seek them yourself. They're there. I promise. And if you keep at it, you'll not only find them. You'll learn more about how to do logical geometry.

Career Advancement: What's the Best Path?

Okay. We've looked at four maps so far—[1] the prototype map (the Mind of Scientific Genius), [2] the map of process and outcome (Being and Doing), [3] the map of Geographic Direction, and [4] the map of Arithmetic Operations.

In all four cases, we mapped theoretical concepts, albeit, the map of Being and Doing can guide a real life practice as well—meditation. But what about a strictly pragmatic map, such as a map to guide life decisions? Can logical geometry scientifically map such a thing? Is this even possible?

The following is an example of just such a map. My friend Joe and I made this map one day over lunch. Admittedly, that lunch was a long lunch—something like eight hours. Once we started, we just could not stop.

At the time, Joe was a SCORE volunteer (Service Corps Of Retired Executives) and had said he wanted to pick my brain. What he wanted to know was how to tell if someone is ready to begin—or advance—in a career.

The Career Advancement Map: the Back Story

We began our conversation that day reminiscing about the time when Joe was my manager. An applicant had lied on his resume, and Joe had hired an incompetent person.

Could he have known, he asked?

Joe then went on to describe some cases from his work at SCORE. In one case, two women asked Joe if they should invest in a massage franchise. They were about to invest their entire retirements.

In another, a disabled, old-age-home bingo-caller wanted to open a hot dog stand.

And in another, an unemployed mortician wanted to become a big city librarian.

And no, I'm not making these stories up.

Eight hours and many stories later, to my surprise, we'd created a map. You'll find this map on the next page. In it, you'll see a number of things, including the answers to Joe's career counseling questions.

You'll also find the map's two axis-defining questions which together narrow down all career advancement choices to just four choices. You'll also find definitions for the terms I've used.

What did I learn from this map?

For one thing, it clearly explains why even well-educated people need experience.

For another, it reveals that going back to school can, at times, make things worse.

Most interesting of all though is the idea that in theory, your career can only be in one of four states.

- Uneducated / Bad Fit
- Uneducated / Good Fit
- Educated / Bad Fit
- Educated / Good Fit.

Here I define "educated" as *having the knowledge to function well in a career*—in effect, having a formal education. And I define "fit" as *using the knowledge you already have to improve your performance in this career*—in effect, using what you know to pragmatically better yourself.

How to Time Career Advancement
(The Career Counselor's Guide Map)
(© 2010 Steven Paglierani The Center for Emergence)

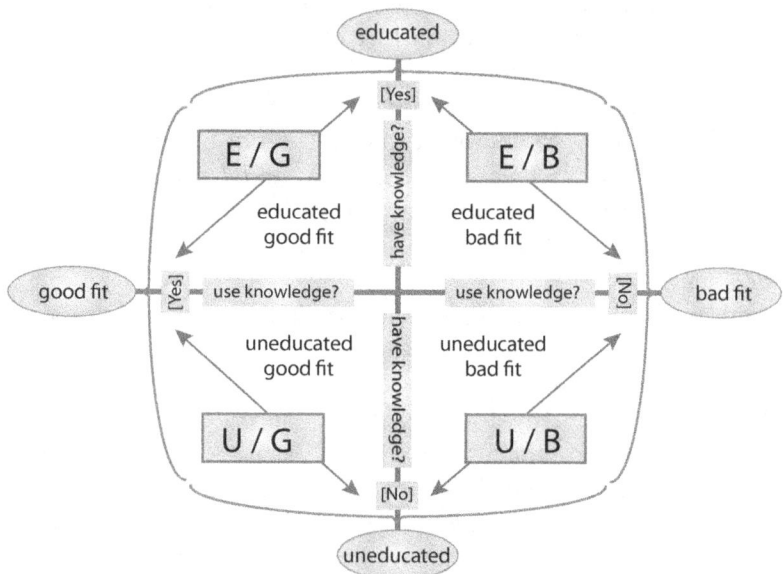

This Map's Two Terms Defined
educated = having the knowledge to function well in a career
fit = using the knowledge you already have to improve your performance in this career

This Map's Two Tipping Point Based Questions
Does this person have the knowledge to function well in this career?
Does this person use the knowledge he or she already has to improve her performance?

The Map's Four Truths

U / B = This person does not have the knowledge and does not try to use what he or she already knows to create a better fit

U / G = This person does not have the knowledge but does try to use what he or she already knows to create a better fit

E / B = This person does have the knowledge but does not try to use what he or she already knows to create a better fit

E / G = This person does have the knowledge and does try to use what he or she already knows to create a better fit

Career Advancement: How Do We Test the Guide Map?

Of course, scientifically mapping a theory is one thing. But this map claims to be a real world guide. So at this point Joe and I began to test this map. To do this, we began to plug some of Joe's real world stories into this map.

We started with his observations of the two women who wanted to start a massage franchise. Here, I asked Joe a number of questions—things like "did these two women know anything about doing massage," and, "did they have any experience running a business?" I then used the two tipping-point based questions—*do you have the knowledge to do this career* and *are you using the knowledge you already have to create a better fit*—to place these observations into the map.

The result? These two women were clearly U / Gs. They knew nothing about running a business, let alone a massage business. But they were doing all they could with the knowledge they had to better their chances.

What kind of advice should Joe have offered them? To see, I'll need to explain the map to you. In particular, I need to explain the four steps that an ideal career must take. Here, each step has a best timing and a worst.

Career Advancement: What's the Best Next Step?

Did you notice I just referred to the "ideal" career advancement path? Know I say this as, in theory, there is a best way to advance though a career. Ideally, you would go from U / B, to U / G, to E / B, to E / G. In the real world, however, this order can—and often does—vary. Before we explore these variations though, we first need to go through the theoretically, ideal path.

This path begins in the lower right quadrant, with the U / Bs—the folks who lack education and don't use what they already know to create a better fit. What should these folks do to advance in their careers?

- **U / Bs—Uneducated / Bad Fit**

To me, this quadrant reveals one of the more surprising things in this map—the idea that going back to school is wrong for some people. Why? Because if you're in this state, then something is keeping you from using what you already know. Thus learning more about what you're not using would just waste your money and time.

What should you do then?

You should focus on finding the source of your stuckness. Here talking to a therapist might help. Perhaps something has been killing your confidence. Perhaps you're afraid to make a mistake.

Then again, perhaps you feel ashamed of your lack of education. Or maybe the idea of taking on more responsibility scares you. Or maybe you see successful people as shallow and insincere. Or maybe you've wanted to choose a career that everyone says is foolish.

Whatever the case, know that no one is born with these feelings. Babies routinely overcome far more difficult things. The point is, feelings like these often point to a wound. Which is why going back to school at this point would only make things worse. Acquiring more knowledge would only confuse you and lead to more feelings of failure.

And after going to therapy?

For many people, working with a job coach can help. We can all benefit from someone teaching us how to better manage and organize our time. Some U / Bs don't use what they know because they can't motivate themselves. Others do feel motivated—but just can't manage their lives well enough to act on this motivation. And others simply don't know where or how to begin.

Then again, some U / Bs have resentments or biases which handicap their efforts. As I mentioned above, some U / Bs resent people who have more education. Others think only physically hard work counts as work—that office jobs don't matter. I myself used to feel this way.

In addition, some U / Bs would like to learn to work with their hands. But they believe, they can't—that they're incapable. Others hate people with management titles, or see folks who work hard as brown-nosers. And others think the world is just against them.

Whatever the case, the goal here will always be the same—to improve your ability to use what you already know. No matter whom you turn to, then—whether psychotherapist, job coach, or spiritual counselor—regardless, you'll need to address these concerns before taking the next step.

Finally, how will you know when you're ready to take the next step? Here the test is actually quite simple. Are you curious—or are you certain? If you feel certain you are ready, then realize, you are not. However, if you realize you're not ready, but feel curious as to how things are done in this career—and if this curiosity has led to your seeking this information all on your own—then you are likely ready to move on to the next step.

What's next?

- **U / Gs—Uneducated / Good Fit**

Like U / Bs, U / Gs also lack education. But U / Gs look for times when they can adapt what they know to what they're learning. This means going back to school will be their best next step—as the main thing holding them back will be their lack of education.

If this is you, how should you decide what to study? To be honest, addressing this topic is far beyond the space we have here. This said, if you speak with people already employed in your chosen career, this can broaden your choices. This includes even people who never went back to school, but wish they had.

In addition, it can be helpful to speak to professionals who specialize in career advancement. This includes college guidance counselors, company HR reps, career counselors, and so on. Here, the thing to keep in mind is that all U / Gs want to be in situations where they can use what they're learning. In effect, you're going back to school at a time when it has a purpose. Moreover, if you invest time in choosing what to study, you'll be well on your way to the next step on the career advancement ladder.

- **E / Bs—Educated / Bad Fit**

What are E / Bs like? For the most part, they've completed their formal educations. At the same time, they are also just starting out and likely stuck at the bottom of the ladder. Thus while in theory, these folks have all they need to do well in their careers, in reality, they often lack the experience to use this knowledge. Sadly, this combination can lead to bad attitudes—over confidence, arrogance, or naïveté, to name a few.

Know this is where the "B" in the E / B designation comes from. This is not to say E / Bs are bad people—or that they're bad at what they do. Rather it's more that they don't realize—or have a hard time admitting—that there's much they have yet to learn. Indeed, at times, this post-educational blindness can lead to clashes with more-experienced folks, especially with those who have less formal education than they do. It can also lead them to have problems with bosses they see as being "old school"—people who finished their formal educations long ago.

The problem of course is that, left unaddressed, these attitudes can lead to feeling chronic dissatisfaction. Ultimately, these folks may even conclude they've chosen the wrong career and quit. Whatever the source of their attitudes though, E / Bs must address these issues. Otherwise, they'll have a hard time advancing in their careers.

Here again—talk therapy, job coaches, and spiritual counselors can all be useful. Practices which develop self awareness and self discipline—such as meditation and martial arts—can also help. Whatever method they choose though, the goal should always be the same. They need to learn to see the good in getting experience, and round off some of their rough edges.

Do this and they'll steadily advance toward the fourth and final step—the E / G step.

- **E / Gs—Educated / Good Fit**

Finally, we come to the folks who have acquired both the requisite, formal education and the practical experience. These folks have a good fit, both theoretically and in the real world. For the most part, this is because they're aware that learning is a lifelong process. Thus most E / Gs continuously look to gain more education and experience.

What advice do you give someone who is at this stage?

For one thing, to keep doing what they're already doing—learning new things, then practicing what they've learned. For another, to take classes which promote creativity—activities which get them to think outside the box. The main thing they'll need to do, however, will be to find ways to share what they've experienced. Nothing improves an E / G's skills more than teaching others what they've learned. Indeed, when E / Gs step into the teacher's role, they really come into their own. And seeing others benefit from their life experiences can be the best reward of all.

Is it really possible to feel this good about a career? All I can say is, it has been for me. I also know other E / Gs, and they feel this way too. Indeed, most of them tell me they so love what they do that they can't imagine retiring. This is exactly how I feel. And if you let this career map guide your career advancement, then you can feel this way too.

Career Advancement: What's the Real World Path?

As I mentioned, the career advancement path I've just outlined is the *theoretically ideal* path—going from U / B to U / G to E / B to E / G. In real life, though, career advancement can—and often does—deviate from this order. For instance, technology may make a leap and leave you behind. Or a new wiz kid may join your firm and marry the boss's daughter.

At times like these, E / Gs can regress to being E / Bs, or even U / Gs. If this happens to you, then you'll need to reassess the state of your life and career. To do this, you should use the career map's questions to point you to your best, next step. For instance, U / Bs and E / Bs will need to work on their ability to use what they know. And U / Gs and E / Gs will need to look into getting more education.

Then there's the idea that, in the real world, career advancement can be stressful. Again, the solution may require professional help. For instance, when facing a personal crisis, an E / G may regress to an E / B. Perhaps this person's parents have grown old and are facing health problems. Perhaps this person is facing a health problem of his or her own. Indeed, even seemingly positive things like having a baby or buying a new home can provoke these kinds of career reversals.

Whatever the case, the thing to realize is this. Things like helping aging parents and raising a child are both, in and of themselves, careers. This means you'll need to apply the career map's questions to these second careers as well.

Yet another situation which could derail a career would be the feeling of being successful for too long. Even good things can become boring if you do them long enough. This means, if you both love and complain about your job—or about your boss or the company you work for—then it may be time to reevaluate your career and make some changes.

Realize the solution to this kind of unhappiness lies largely in addressing problems in your attitude. Focus on your attitude and your life will improve greatly. This holds true *even if the things you complain about are real*. There's no situation a bad attitude won't make worse.

Lastly, recall my original point for telling you these things. They're all examples of how a single quadtinuum can simplify even complex, real world problems. In truth, there's no career-advancement problem this map can't improve. And considering that going to college is a career, you could even use this map to plan your advances through school.

One map. Unlimited possibilities. Something to consider.

What If You Can't Find a Map's Two Questions?

In a moment, I'll briefly outline the entire, new scientific method. This method begins with something I've yet to describe, a sequence I call the "Cartesian process." This process gets you ready to progress through the six geometries. It also supplies the raw data for the two questions. And if you've followed my initial map-making discussions, then you realize everything hangs on being able to find these two questions.

What happens if you can't come up with these two questions? To be honest, I've had this happen a lot. For example, take the four tastes—sweet, sour, salty, and bitter. Intuitively, I sense these four things are complementary opposites. I'm also fairly certain they account for all tastes, at least, the natural ones. And yes, I know, some say there are five tastes—some say, even more. But without a map, I cannot know—let alone whether these four tastes are answers, classes of discovery, natural states, or myths.

I also have yet to find questions for the four forward routes—over, under, around, and through. Ditto for the four computer-mouse motions—up, down, left, and right. Then there are the four dichotomies of Myers Briggs personality test—I / E, S / N, T / F, and J / P. As well as the Four Noble Truths of Buddhism, and the four New Testament gospels.

How about the four horsemen of the apocalypse—are there relationships hidden there? And what about the four seasons, the four elements, and the four playing card suits? What would mapping these things reveal?

Then we have the four basic states of matter (liquid, solid, gas, plasma), the four blood types (A, B, AB, O), and the four bases from which all DNA forms (A, G, C, T).

Can you imagine the connections we might discover if we could map these things?

For example, would we find that Buddhism, the four gospels, and the Myers / Briggs tests are just different incarnations of the same personality theory?

Would we discover that the basic states of matter, DNA, and blood types are all stops on one progression, from inert to alive?

Perhaps certain combinations of blood types, DNA, and exposure to states of matter correlate to people's religious beliefs?

And perhaps, the Four Noble Truths are positive incarnations of the four horsemen of the apocalypse?

So How Could We Benefit From Logical Geometry?

Are you beginning to see how logical geometry could help you unravel the world's, scientific mysteries? Can you also see how it would promote cooperation between you and your fellow scientists? Moreover, can you also see how logical geometry could limit the scope of your searches to valid journeys, making you a more thorough, more honest, and more evidence based scientist? Most important of all, can you imagine a scientific method which guarantees discoveries? How amazing would it be to have this. And yes, it's one thing to gain theoretical knowledge—quite another to gain real world skills.

Fortunately, constellated science can deliver both.

So at this point, are you at least beginning to feel curious?

The Old & New Sciences: What's Different?

Before I close this chapter, I'd like to briefly contrast and compare the new scientific method with the old. The thing is, as I mentioned a moment ago, I've yet to introduce you to some parts of constellated science's method. For example, in this brief comparison, I'll refer to the "Cartesian process." Know I'll introduce you to this process in the next chapter.

As for how I'll make this comparison, I'll begin by briefly describing each method. Here, each section will include a series of four actions and four goals. We'll start with the current method.

What are the actions and goals of the current scientific method?

The Current Scientific Method
- Formulate a predictive hypothesis which disqualifies all but the desired observations (action). This focuses the work (goal).
- Design and execute cause-and-effect experiments (action) which turn these observations into logically sensible data (goal).
- Filter and arrange this data (action) until you arrive at a logically sensible outcome (goal).
- Contrast and compare this outcome with the hypothesis (action) in order to arrive at a logically sensible conclusion (goal).

And the actions and goals of constellated science?

The Constellated Science Method
- Use the Cartesian process—slate clearing, fact gathering, experimenting, and pattern seeking—to open and direct your mind (action). This insures you'll value intuitions, facts, experiments, and concepts equally—and guarantees you'll omit none of what you observe (goal).
- Use geometries One through Four to begin to gather and arrange a geometrically logical progression of evidence (action). This insures your observations will be focused, comprehensive, scientific, and relevant (goal).
- Use geometries Five and Six to geometrically arrange and correlate this evidence (action). This guarantees you'll discover connections and relationships within this evidence (goal).
- Constellate these interconnections with those in all other maps (action). This functions as both a test and a proof for the validity of your work. If you fail to find these parallels, this indicates an inherent incompleteness. Conversely, if you find these parallels, this validates and integrates your discoveries into science's entire body of work (goal).

Are you beginning to see how these two methods differ from each other? One method progressively separates you—and your work—from the rest of the world. The other progressively integrates you—and your work—into the world. Moreover, if we now contrast and compare these two methods, we arrive at the following, initial conclusions.

Difference 1: How You Decide Which Evidence to Consider

The current scientific method begins with a predictive hypothesis—an educated guess—a proposed answer. This answer assumes abstract, cause-and-effect relationships underlie the nature of all things. It also assumes this logic can be used to predict a thing's nature—moreover, that researchers can use these predictions to decide which evidence is relevant.

Unfortunately, doing this eliminates most evidence *before* the observation process begins. This is like deciding to leave out numbers in an addition problem—and believing you can still get a correct answer. In effect, these omissions are the current method's "emperor's new clothes." *They significantly reduce a scientist's chances to make discoveries. But no one mentions this.*

Constellated science begins by asking a non-predictive question. A "hypothesis"—a predictive new line of questioning—can appear only in "what's next" points in the process. This assumes all things in the natural world interact—and that fractal, rather than logical, patterns are the only predictive part of nature. This means, like an addition problem wherein you must include all the numbers, in constellated science, you must consider everything that's been observed. Indeed, it sees any method which selectively excludes observations as unscientific. Thus, the first principle in constellated science is, all observations must be accounted for. Doing this *significantly expands a scientist's chances for making discoveries.*

Difference 2: How You Gather This Evidence

As I said, the current scientific method assumes logical, cause-and-effect relationships underlie all natural things. This explains why so many scientists refuse to consider things like emergent properties and intuition (for more on this, see Book I, chapter 3). It also explains why scientists try to avoid experiments which result in emergent properties and intuitive contradictions. They hope eliminating this evidence will result in repeatable, reliable, cause-and-effect outcomes. And many times, it does.

The thing is, by doing this, they treat all nonlinear, illogical, and physically-unmeasurable evidence as anomalies or annoyances. This severely limits the value of their experiments. Any time you patently discard part of a problem, you guarantee your results will be incomplete. This incompleteness then leads to results which cannot translate well to the real world. Ergo the present scientific method's oft quoted flaw.

With constellated science, scientists assume all things have a fractal nature. This nature includes emergent properties which logic could never predict. This is why constellated science see logical geometry as the only valid means for collecting observations. Only logical geometry guarantees no evidence—no matter how seemingly irrelevant—can ever be omitted.

This wholeness then insures the results of this method translate well to the real world. Moreover, because this process by nature progressively expands the observed field of relevant evidence, the longer it continues, the more likely it becomes that you'll make valid scientific discoveries.

Difference 3: How You Arrange The Evidence You've Gathered

With the current scientific method, scientists assume that beneath all real world, messy roundness lies an abstractly statistical nature. This is why—when these scientists arrange their observations—they see nothing wrong with omitting and devaluing any data which does not fit into a statistically logical schema. Indeed, these folks actually see discarding statistically-insignificant evidence as an essential part of the discovery process. Sadly this assumption reduces their chances for making discoveries even more.

With constellated science, scientists assume that real world messiness *is* the true nature of the world. Thus they look to arrange what they observe into naturally occurring patterns. Moreover, since logical geometry always underlies these naturally occurring patterns, if something does not fit, it does not get discarded. Rather this lack of fit is seen as the evidence for either a flaw in the arrangement of the data—or an insufficient number of observations.

The result? Massaging incomplete data so that it fits into patterns is seen as an ethical violation. To be considered good science, a scientist must discover patterns which address all the data. If not, then she or he must keep looking until such patterns emerge. Moreover, constellated science sees times wherein all the data does not fit as opportunities to make more discoveries. Thus with constellated science, the possibilities for discovery increase with each new observation.

Difference 4: How You Arrive At Your Conclusions

With the current scientific method, scientists reach conclusions by contrasting and comparing the filtered evidence to their working hypothesis. This filtering assumes scientific discoveries must, by nature, be statistically logical. This effectively rules out any and all conclusions which don't statistically correlate to the hypothesis. This severely limits their possibilities for discovery.

With constellated science, scientists reach conclusions by contrasting and comparing what they've found to all other findings. Doing this assumes all things in the natural world connect and interact. Here, any scientific discovery which does not directly correlate to all other known discoveries is considered incomplete or flawed. This raises the bar for all branches of science, because it rules out conclusions which do not constellate to all other branches. And because the natural world is infinitely complex, this criteria effectively raises the possibilities for discovery to an unlimited number.

A Few Closing Thoughts About Two Methods

By now, I hope you're beginning to see the differences between the two scientific methods, starting with the two main differences—that the current method deliberately omits evidence in order to come to conclusions, and it forbids certain topics from being explored. Whereas constellated science requires that you omit nothing, neither evidence nor topics.

In large part, this explains why many otherwise excellent scientists fail to make discoveries. The present method so narrows their focus that most discoveries occur accidentally. Does this claim sound crazy? Then try reading Sebastian Seung's book, *Connectome* (2012). In this book, Seung—a noted neuroscientist—proposes what he calls the efficient science hypothesis (ESH). This hypothesis claims that "no fair and certain method of doing science" will ever exist.

In effect, what Sung is saying here is that there will never be a scientific method which guarantees discoveries. Seung then goes on to say that better tools can improve the odds—but that there is no way to ensure breakthroughs.

The thing is, even a casual read of Seung's book reveals he's a genius. Moreover his level of honesty about the current state of science is beyond rare. I applaud him.

I also agree with his assessment of the current scientific method—its discoveries are largely accidental. However, to be blunt, Seung's own hypothesis—that no method can guarantee discoveries—is flat out wrong. Of course, to evaluate this claim, you'll first need to free yourself from the confines of hypothesis-driven science. How? By focusing your initial explorations on creating a logically geometric map, rather than on creating a hypothesis.

No coincidence, Seung himself points to the flaws in using a hypothesis. Considering Seung himself somehow misses the shortcomings in his own hypothesis, I find this especially ironic. For instance, consider the core concept of connectomics—that intelligence lives in the patterns of connections. Admittedly, this concept is what drew me to his work.

This said, to me, it's sad that such a brilliant man can have missed the brilliance of his own work. I say this as he somehow overlooks the idea that what he's applied to the brain needs to be applied to all of science. The structure of our brains and our intelligence simply mirrors the underlying structure of the whole world. And logical geometry reveals the structure underlying them both.

As for the specifics of the new method—the four-part, Map-Making process and the logical geometry I've outlined in this chapter comprise a good part of it. Know I'll present additional concepts throughout the rest of the book which further define and support these claims.

For example, in the next section, we'll look at how a geometric figure known as the Möbius Strip can prepare and open your mind. To do this, we'll explore four topics—the nature of sleep, the nature of weight loss, the nature of deafness, and the nature of cancer. Can we discover anything new? We're about to find out. Know that at this point, I honestly have no idea what we'll find. I swear.

Then, in a future chapter, I'll introduce you to an algebra which may hold the key to deciphering Seung's connectome. In this case, I've used the six geometries to map the entire structure of the mind. No surprise, to do this, I needed to expand logical geometry to x, y, and z space. In other words, this map is three-dimensional.

Later on, I'll take you through some of the things I'm currently working on—the nature of creativity, the nature of learning and education, the nature of consciousness and the mind, and the nature of romantic attraction. And at some point, we'll even look at how we might map a whole profession—talk therapy—as well as why we have such a hard time trusting even things as logically elegant as the six geometries.

What makes us so doubtful and defensive when it comes to new ideas? And what makes scientists so quick to call "free thinkers," pseudoscientists? More important, why do so many of us feel we're not qualified to question scientific claims? Moreover, what makes me think I am?

As Yoda might say, "All this and more. Find out, you will."

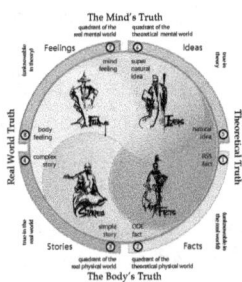

Section One - End Notes (Ch. 7 thru 9)

Afterthoughts & Resources

Notes Written in the Margins of Chapters 7, 8, & 9

How This Book Parallels Descartes' Discourse on Method

In 1637, Descartes raised the bar for science when he published his *Discourse on the Method for Directing One's Reason Well and Searching for Truth in the Sciences*. In this book, Descartes describes his revolutionary scientific method, along with his most famous legacy; *I think, therefore I am* ("Je pense, donc je suis").

I mention Descartes and his book for three reasons. The first is that his four step method roughly parallels a significant aspect of the method I am presenting in this book; the part I've named the Cartesian process. Of course, where Descartes' four steps are his entire method, in Constellated Science, the Cartesian process is just the beginning.

That even the order of this process's four steps is the same never ceases to amaze me. I became familiar with Descartes' four steps only after discovering these steps on my own. Nonetheless, when I wrote Book I, I felt it important to credit Descartes. Were it not for him, I probably wouldn't have come to my own version of his work.

Crediting Descartes then is my first reason. Obviously, this is important. But the second reason is far more relevant to the present book.

Here I'm referring to the format Descartes uses to arrange his chapters. He begins with a preface in which he outlines his method. Three sections then follow which document his discoveries on optics, meteorology, and geometry.

In this book, I'll follow a similar path. I'll first outline my method. Then I'll describe the discoveries I've made using this method.

The third and final reason I mention Descartes involves yet another curious parallel between his work and mine. This time, I'm referring to the idea that the sequence of logical geometry I've just introduced could not exist were it not for Descartes' geometry—Cartesian Coordinates. Moreover, similar to the way I see my four steps as refining his four step method, I also see logical geometry as an extension of his analytic geometry. In other words, where his geometry marries algebra to geometry, logical geometry marries algebra and geometry to logical words.

I mention these things as I feel deeply indebted to Descartes. As I said, without him, it's likely none of what I'm presenting here would exist.

Cartesian Coordinates & Quadtinuums: a Few Last Words

As I've just said, I've repeatedly been amazed by the many parallels between Descartes' work and mine. This includes my last example, wherein it occurred to me that the prototype map is a variation on Descartes' coordinate system—his combination of algebra and geometry. Know we'll be looking at this idea in depth in a future chapter, when I introduce you to an algebra which uses just four simple formulas to describe the essence of human consciousness. And as heady as this may sound, these formulas may just change your life. In truth, they have the power to reveal many previously hidden truths—everything from how getting startled permanently closes minds to why people get addicted to crack more easily than to marijuana. You can even use these formulas to scientifically plot human experiences like grief and human emotion. More interesting stuff.

Is Shelf Life (unchange) a Test for Toxicity?

One way to determine the real world toxicity of foods is to look at their shelf life. Here toxicity is directly related to the length of a thing's usable lifetime. Ironically, in theory, long shelf life is also a test for how well something is made. Strangely, it's also one of the main ways scientists test for truth.

The thing is, while long shelf life is often seen as a positive quality in non living things, in the real world, *a short shelf life* is often a better gauge of quality. Yet another example of the good in bad.

For instance, say we were talking about a legendary Twinkie. If you left it out on a counter for a year and saw no change, would you eat it? Or say we were talking about freshly cut roses. If they looked the same at the end of a month, would you say these roses were better roses?

What if you had a baby who never changed and stayed perfectly new? Would this be a test of the baby's goodness? Or what about a husband who never aged? Would this be a good thing?

The point is, as I've said, for a scientific method to be considered true, it must account for both the positive and negative sides of change—and the lack thereof. For instance, where there is no change in the real world, there is no quality of life and vice versa. Thus while in theory it's great to have something last forever—including the answer to a question—in the real world, this can often be a bad thing.

No coincidence, lasting forever is a single point.

Do Continuums Create Techniques or Tools?

Some continuums create tools. Some create techniques. And some create neither.

They create techniques if they have a finite number of visible uses (a linear continuum).

They create tools if they have an infinite number of visible uses (a fractal continuum).

And they create neither if they have no visible uses (an unrecognizable geometry).

On "Being" Right, and "Doing" Wrong

If *being* refers to witnessing change and *doing* to witnessing no change, then we can't *be* wrong and we can't *do* right. We can only *be* right and *do* wrong. Here, in order to judge something as a *wrong*, we'd need to freeze a moment in time. And in order to judge something as *right*, we'd need to see how this thing affects what is currently changing.

The point?

Being right is a real world condition which is always temporary. Whereas *doing wrong* is a theoretical condition which is always permanent.

At least, in theory.

On These Maps as an Artificial Intelligence

When my editor read my claim that the prototype map is an artificial intelligence, I realized from her comments that I'd once again stepped in shit. People have written whole books in their attempts to define artificial intelligence. Yet I cavalierly boasted this map of the mind contains artificial intelligence without first defining this term.

The thing is, when you understand the six geometries, you realize where this is coming from. Taken together, they form the core of all intelligence, visual and otherwise.

Oops, there I go again. Let me try again.

In the final chapter of this book, I'll take you through a few examples of how the new scientific method could explore as yet unsolved, scientific mysteries. In each case, just placing a few words into the map led to some new discoveries. Of course, knowing which words to plug into a map still requires an intelligent being. But this in no way negates my claim. Anything which embodies all aspects of IQ is smart as hell.

And the idea that these maps require some human intervention? Well, don't we all?

On Previous Theories of Scientific Reality

One thing I've yet to mention is a problem all scientific methods must address. By this, I mean the problem of whether this method's discoveries describe a literal reality—or a representation of this literal reality; an approximation of this reality. For instance, for Percy Williams Bridgman, the only reality science can trust is operationalized functions and their resultant facts. Models alone aren't enough. Whereas, for folks like Bertram Russel and Ludwig Wittgenstein, the only reality science should trust is analytic logic and reason. Facts alone aren't enough.

Fortunately, for the uses of constellated science, this question has been addressed. To be true science, any claimed reality must includes facts, stories, ideas, and feelings. More important, it must result in an ongoing stream of new discoveries, as well as an unlimited ability to predict real world outcomes. After all, in the real world, reality is never done. The real world is always changing. At least in theory, it is.

Logical Geometry as a Formal Mathematics

Finally, for those to whom it is important that I give a formal explanation for my claim that logical geometry is a new mathematics, I offer the following brief description.

Mathematics is generally defined as the abstract science of number, space, and time. Moreover, as we've discussed in this chapter, logical geometry employs all three plus logic. Here, logical geometry employs a formalized progression of six, phylogenically symmetrical shapes. The axis points in these shapes then function like placeholders for words and phrases.

A set of rules then guides the data entry process, including that all entries must be logically and geometrically analogous, including that they must employ the same logical symmetry, both internally and externally.

Why require all aspects of this mathematics to be logically and geometrically symmetrical? Because in nature, asymmetry is the norm. Faces. Apples. Clusters of stars. In each case, true symmetry is rare. This means, when symmetry does occur, it makes this pattern scientifically significant. Moreover, when this symmetry nests within layers of self-similar symmetry, the significance of these patterns is beyond rare.

In logical geometry, the basic symmetry is the numeric progression which births all six forms—the numeric sequence; 1, 2, 4. Thus, all things start as a single thing. This single thing then bifurcates into a pair of complementary opposites. Then each of these two opposites bifurcate into two more complementary opposites as well.

Again, statistically, the results of this sequence are beyond calculation—in each case, a single set of four points, each of which is the complementary opposite to the other three. That all bifurcations must result from "tipping-point based math" makes it even more rare. Why? This advanced counting math employs *deliberately induced bifurcations in a closed system for the purpose of measuring non linear processes*. This makes tipping-point based math the only counting math capable of measuring real world things with 100% accuracy.

What makes logical geometry a formal math then? Several things.

[1] It is an interdependent, abstract science of number, space, time, and logic, wherein no ad hoc expressions are allowed.

[2] All expressions must be logically and geometrically analogous to all others.

[3] All expressions must be both internally and externally self-similar, as well as logically symmetrical.

[4] All expressions must be capable of measuring real world quantities and qualities with 100% accuracy.

[5] All expressions must be universal in nature (meta-physical).

IMPORTANT: Must You Parallel All Points to Make a Map?

During the writing of this section, my editor repeatedly asked me a question. How could I place things in the example maps without knowing how to parallel every aspect in these new maps to every aspect of the prototype map? For example, she asked how things like *east* and *west* in the map of direction parallel *change* and *unchange* in the prototype map. At which point, I realized I had unknowingly overlooked this in the chapter.

Know it all comes down to whether you're mapping spatial concepts or non-spatial concepts.

What I mean is, when a map employs an already existing set of four *spatially related* concepts—such as in the map of the four directions—

paralleling a single point in this group allows the rest to be positioned. Here, once north is placed in a map, the placement of the other three directions is fixed. Similarly, with left , right, up, and down. Once left is placed in a map, the placement of up, down, and right are fixed as well.

When concepts which are not spatially related though—such as with the four arithmetic operations—the placement process tends to be a bit more difficult. To place things in these maps, you must first parallel at least one point on each axis. For instance, in the map of arithmetic, grouping addition with multiplication (results in an increase) gave me one point, *but no placement*. Grouping addition with subtraction (the entire process is visible) then gave me the second. At which point, I knew how to place things in the map.

Finally, as I briefly mentioned in chapter 9, constellated science employs four processes. The as yet to be described fourth process—the Constellating process—completes this paralleling.

On Fractals as Nature's Linearity

My editor asked why I wrote that fractals have two halves—one half recognizable, the other unrecognizable. To see, forget about sizes and areas and focus only on qualities. One half of a fractal's qualities involve the quality of recognizability. The other half—the quality of unrecognizability.

Resources for Section 1—Logical Geometry

On Geometry as a Path to Truth

If Pythagoras was alive, I imagine he'd love this chapter. At least the parts which say that geometry holds the key to purposely making discoveries. Indeed this idea is one of the few ideas that the four protoscientists (the materialist, the empiricist, the rationalist, and the spiritual scientist) can agree on. At least, in theory. This said, here are a few of the sources I consulted during this chapter.

Giaquinto, Marcus M. (2007). *Visual Thinking in Mathematics*. New York: Oxford University Press. (A difficult read, but quite thought provoking.)

Kline, Morris. (2009). *Mathematics : The Loss of Certainty* (Barnes & Noble Rediscovers Series). New York: Barnes & Noble. (The stuff on Pythagorean tetractys and fournesses as the true essence of nature floored me.)

Kahn, Charles H. (2001). *Pythagoras and the Pythagoreans, A Brief History*. Indianapolis, IN: Hackett Publishing Co.

Tubbs, Robert. (2008). *What Is a Number?: Mathematical Concepts and Their Origins.* Baltimore, Maryland: Johns Hopkins University Press.

Lundy, Miranda, and D. Sutton, A. Ashton, Ja. Martineau, & Jo. Martineau. (2010) *Quadrivium: The Four Classical Liberal Arts of Number, Geometry, Music, and Cosmology.* New York: Bloomsbury Publishing. (As you might imagine, I love finding already existing-parallels to my work. Here is yet another ancient quadtinuum: Number as Number [arithmetic], Number as Space [geometry], Number as Time [harmony], Number as Space and Time [astronomy]. Brilliantly presented and summarized.)

On Shelf Life as a Real World Test for Quality

Since durability can be both a test for quality and the lack thereof, I find charts of shelf lives amazing, whether they be of foods, chemical formulas, or atomic states. And as these two sources show—with food, durability confers the later—whereas with coatings, it confers the former.

Man, Dominic. (2002). *Shelf Life: Food Industry Briefing Series.* New York: Wiley, John & Sons.

Hardcastle III, Henry K. (2010). *Variables, Methods and Philosophies Considered in Coatings Durability.* Chicago, IL. retrieved 4/9/10 from http://www.atlas-mts.com/en/client_education/weathering_library/service_life_prediction_and_numerical_weathering_test_methods/variables_methods_and_philosophies_considered_in_coatings_durability_44917.shtml.

(This sidebar is representative of the philosophical position that longer shelf lives equate to quality. It's also an interesting article.)

On the Efficacy of St. John's Wort

Here I list the two studies I referred to in this chapter. Moreover, to save you the time of searching these studies out, I've included detailed abstracts of both.

§

A Study Which Claims St. John's Wort is Ineffective

National Center for Complementary and Alternative Medicine (NCCAM), National Institute of Mental Health, Office of Dietary Supplements (2002). S*tudy Shows St. John's Wort Ineffective for Major Depression of Moderate Severity.* http://nccam.nih.gov/news/2002/stjohnswort/pressrelease.htm (The National Center for Complementary and Alternative Medicine (NCCAM) is the Federal Government's lead agency for scientific research on complementary and alternative medicine (CAM).

JAMA. 2002 Apr 10;287 (14):1807-14. Retrieved Feb 2, 2012 from http://www.ncbi.nlm.nih.gov/pubmed/11939866?dopt=Abstract.

Effect of Hypericum perforatum (St John's wort) in major depressive disorder: a randomized controlled trial (Hypericum Depression Trial Study Group).

Abstract:

CONTEXT:

Extracts of Hypericum perforatum (St John's wort) are widely used for the treatment of depression of varying severity. Their efficacy in major depressive disorder, however, has not been conclusively demonstrated.

OBJECTIVE:

To test the efficacy and safety of a well-characterized H perforatum extract (LI-160) in major depressive disorder.

DESIGN AND SETTING:

Double-blind, randomized, placebo-controlled trial conducted in 12 academic and community psychiatric research clinics in the United States.

PARTICIPANTS:

Adult outpatients (n = 340) recruited between December 1998 and June 2000 with major depression and a baseline total score on the Hamilton Depression Scale (HAM-D) of at least 20.

INTERVENTIONS:

Patients were randomly assigned to receive H perforatum, placebo, or sertraline (as an active comparator) for 8 weeks. Based on clinical response, the daily dose of H perforatum could range from 900 to 1500 mg and that of sertraline from 50 to 100 mg. Responders at week 8 could continue blinded treatment for another 18 weeks.

MAIN OUTCOME MEASURES:

Change in the HAM-D total score from baseline to 8 weeks; rates of full response, determined by the HAM-D and Clinical Global Impressions (CGI) scores.

RESULTS:

On the 2 primary outcome measures, neither sertraline nor H perforatum was significantly different from placebo. The random regression parameter estimate for mean (SE) change in HAM-D total score from baseline to week 8 (with a greater decline indicating more improvement) was -9.20 (0.67) (95% confidence interval [CI], -10.51 to -7.89) for placebo vs -8.68 (0.68) (95% CI, -10.01 to -7.35) for H perforatum (P =.59) and -10.53 (0.72) (95% CI, -11.94 to -9.12) for sertraline (P =.18). Full response occurred in 31.9% of the placebo-treated patients vs 23.9% of the H perforatum-treated patients (P =.21) and 24.8% of sertraline-treated patients (P =.26). Sertraline was better than placebo on the CGI improvement scale (P =.02), which was a secondary measure in this study. Adverse-effect profiles for H perforatum and sertraline differed relative to placebo.

CONCLUSION:
This study fails to support the efficacy of H perforatum in moderately severe major depression. The result may be due to low assay sensitivity of the trial, but the complete absence of trends suggestive of efficacy for H perforatum is noteworthy.

A Study Which Infers St. John's Wort is Effective (given you combine it with light therapy)

Am J Psychiatry. 2006 May;163 (5):805-12. The Can-SAD study: a randomized controlled trial of the effectiveness of light therapy and fluoxetine in patients with winter seasonal affective disorder. Lam RW, Levitt AJ, Levitan RD, Enns MW, Morehouse R, Michalak EE, Tam EM. Source: Mood Disorders Centre, UBC Hospital, 2255 Wesbrook Mall, Vancouver, BC V6T 2A1. r.lam@ubc.ca

Retrieved Feb 2, 2012. http://www.ncbi.nlm.nih.gov/pubmed/16648320

Abstract:
OBJECTIVE:
Light therapy and antidepressants have shown comparable efficacy in separate studies of seasonal affective disorder treatment, but few studies have directly compared the two treatments. This study compared the effectiveness of light therapy and an antidepressant within a single trial.

METHOD:
This double-blind, randomized, controlled trial was conducted in four Canadian centers over three winter seasons. Patients met DSM-IV criteria for major depressive disorder with a seasonal (winter) pattern and had scores > or = 23 on the 24-item Hamilton Depression Rating Scale. After a baseline observation week, eligible patients were randomly assigned to 8 weeks of double-blind treatment with either 1) 10,000-lux light treatment and a placebo capsule, or 2) 100-lux light treatment (placebo light) and fluoxetine, 20 mg/day. Light treatment was applied for 30 minutes/day in the morning with a fluorescent white-light box; placebo light boxes used neutral density filters.

RESULTS:
A total of 96 patients were randomly assigned to a treatment condition. Intent-to-treat analysis showed overall improvement with time, with no differences between treatments. There were also no differences between the light and fluoxetine treatment groups in clinical response rates (67% for each group) or remission rates (50% and 54%, respectively). Post hoc testing found that light-treated patients had greater improvement at 1 week but not at other time points. Fluoxetine was associated with greater treatment-emergent adverse events (agitation, sleep disturbance, palpitations), but both treatments were generally well-tolerated with no differences in overall number of adverse effects.

CONCLUSIONS:

Light treatment showed earlier response onset and lower rate of some adverse events relative to fluoxetine, but there were no other significant differences in outcome between light therapy and antidepressant medication. Although limited by lack of a double-placebo condition, this study supports the effectiveness and tolerability of both treatments for seasonal affective disorder and suggests that other clinical factors, including patient preference, should guide selection of first-line treatment.

The point, of course, is that, St. John's Wort sensitizes you to sunlight. Oddly, the first study failed to take this into account. So much for the so-called scientific method of today. In large part, this error was due to the present method's reliance on the "working hypothesis." Indeed, to this day, I doubt those scientists realize how one's starting assumptions color your outcomes. Are you beginning to see why I'm claiming it's time for a new scientific method? Duh!

On the Brain as a "Connectome"

Sebastian Sung is a brilliantly lucid writer. His analogies are clear; his ideas, interesting. Sadly his medical materialism taints the whole meal. According to Sung, there is no soul. Or anything else, except things which can be physically measured.

Most notably missing from his theories are any references to emergent properties. He also fails to connect the knowable real world to naturally occurring fractal patterns. Sung continues to see logically linear patterns as truth.

Worse yet, nowhere does he mention the idea that the real world measurement problem, let alone the solution—tipping-point based math. To Sung, despite the complexity to the mind, simple counting math is enough. How can such a smart man believe such a thing?

This said—and at the risk of confusing you—I give this book my highest endorsement. Why? Because Sung's explanations of the physiological aspects of neuroscientific research are amazingly clear. For this alone, this book should be required reading for anyone interested in neuroanatomy.

What about the fact that Sung believes the non material aspects of life all reduce to neurons?

Well, geniuses are allowed their biases. And Sung is truly a genius.

Seung, Sebastian. (2012). *Connectome, How the Brain's Wiring Makes Us Who We Are*. Boston: Houghton Mifflin Harcourt.

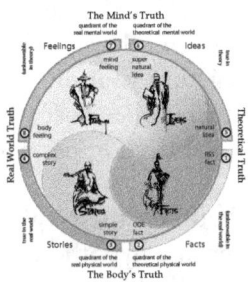

Section 2 - Introduction

The Cartesian Process (what can we learn about ...?

Hypotheses and the Cartesian Process

In Section 1, we took a first look at constellated science and logical geometry. We did this by constellating several areas of interest—*being* and *doing* (meditation), the four directions (geographic movement), the four arithmetic operations (counting math), and the timing of career advancement (a practical guide). In this section we'll dig more deeply into the constellating process. Here, we'll focus on the part I call, the Cartesian process—the part which replaces the scientific hypothesis.

To do this, we'll constellate four more topics—sleep, weight loss, deafness, and cancer. And since constellated science sees connections as its holy grail, we'll again look to discover previously unseen patterns of connections. The thing is, *when possible*, I prefer down-to-earth discussions over confounding technical explanations. Or as Nobel prize winning physicist, Ernest Rutherford, said, "An alleged scientific discovery has no merit unless it can be explained to a barmaid" (G. J. Whitrow, 1973) [1].

The idea behind constellated science is clearly one of those ideas. And to see why I say this, imagine I've handed you a large piece of cardboard with one small hole in it. Now imagine I ask you to hold this cardboard up to a certain part of the night sky—and to center it on a certain star. Now I ask you, "which constellation is this star in?" Could you know?

Contrast this with being asked this same question, but without the cardboard blocking your view. In other words, imagine you can now see the entire night sky. Now picture yourself realizing the mystery star is one of seven in what is arguably the most well known constellation in the northern hemisphere—the Big Dipper. How would this experience differ from the first?

Know the differences between these two experiences are a lot like the differences between scientific hypotheses and the Cartesian process. Scientific hypotheses function like big pieces of cardboard with one small hole in them. The Cartesian process functions like a slow scan of the night sky without the cardboard. With hypotheses, you assume that the part of the sky visible through the hole holds the answer—and that this answer is the goal. Whereas with the Cartesian process, you assume the whole sky holds relevant data—and that the goal is to find patterns in what you see.

Know we'll talk a lot more about these differences throughout this section. But for now, these differences can be boiled down to one pair of opposite actions. A scientific hypothesis is an assumed answer you're trying to prove. The Cartesian process is a quest for questions to explore.

Answers. Questions. Prove. Explore. Can these methods be more different?

Why Sleep, Weight Loss, Deafness, and Cancer?

Why chose these four topics? Because all four have personally touched my life. One of my clients spent ten years working rotating shifts and had not slept normally in years. My mother couldn't gain weight and died of anorexia at age 48. The son of close friends went deaf at six months old. And my father died of prostate cancer—and I still miss him.

Admittedly, I am not a medical doctor. Nor am I trained in the medical arts. So how can I have the gall to think I can discover things about these topics which science does not already know? If constellated science is what I claim it is, then I should be able to make discoveries *even in areas in which I am not formally trained*. So can I? We're about to find out.

Why So Many Groups of Four?

Before we proceed, I first need to offer a bit of advice. Without it, what you're about to read will likely only confuse you. The advice focuses on the idea that constellated science repeatedly refers to groups of four things—four processes, the four steps in the first process, and so on. And unless you're a Pythagorean, all these fours can make things hard to follow.

The simple truth is, for scientific research to succeed, it must limit the scope of its research. Constellated science does this by identifying four complementary opposites which act like the four corner pieces of puzzles. Every puzzle has four corner pieces. So does every one of constellated science's processes—and sections—and steps. If they didn't, they would risk including unrelated material, or make the research incomplete.

The point is, try not to get too hung up on remembering all these sets of fours. In truth, it's far more important to begin to see how this number—four—holds a special place in any scientific endeavor. Including that it can clearly define the full scope of any research.

Constellated Science's "Four Processes"

Before we delve into this section's four topics (there's that number again), I first need to remind you of how the Cartesian process fits into the new method. To begin with, as I mentioned at the end of Chapter 9, the new method consists of a series of four processes—the Cartesian process, the Collecting process, the Map-Making process, and the Constellating process. In the first process—the Cartesian process—you use four steps to open your mind and define your field of vision. And since these four steps roughly parallel the four steps René Descartes outlines in the preface of his 1637 book on scientific method—*Discourse on the Method (for Directing One's Reason Well and Searching for Truth in the Sciences),* I named this process after Descartes.

What about the second process—the Collecting process? As I explained in the previous chapter, you use geometries one through four to begin to organize and define what you find. Then, in the third process, you use geometries five and six to map and explore the connections in what you find. This is the Map-Making process. Finally, in the fourth process, you constellate and parallel these findings to the connections in other maps. This insures nothing is ad hoc. This is the Constellating process.

In the first section of this book, I focused on the second and third processes—on Collecting and Map-Making. In this section, we'll focus mainly on the first process—the Cartesian process. We'll use this process to open and direct our minds, and to guarantee we omit none of what we observe. This insures that we'll employ our whole minds, by valuing scientific intuition, facts, experiments, and concepts equally.

Know I've outlined the Cartesian process's four steps in depth in my previous two books. Admittedly, I worded things differently there as my focus was on finding personal truths—rather than on making scientific discoveries. In case you have yet to read those books though, I'll briefly restate it here.

- **Process One—Step One: the Intuitive Step—Slate Clearing**

A scientific method must lead to *intuitively* true answers. This means you must feel clear about your direction and focus. To accomplish this, you must begin by letting go of all you know—then trust your intuition to guide you to a starting point. And remember. An open mind feels curious, while a closed mind feels certain. So if you feel certain as you begin, your mind is closed.

- **Process One—Step Two: the Material Step—Fact Gathering**

A scientific method must lead to *factually* true answers. To do this, you must divide what you're exploring into discernible pieces, in order that measurements can be taken. In other words, you must gather facts.

- **Process One—Step Three: the Empirical Step—Experimenting**

A scientific method must lead to *empirically* true answers. To do this, you must experiment with the arrangement of these facts until they tell you a meaningfully relevant story.

- **Process One—Step Four: the Rational Step—Pattern Seeking**

A scientific method must lead to *rationally* true answers. You accomplish this step by comprehensively reviewing the previous three steps, until the logic beneath them pleasantly surprises you.

Applied to our first area of exploration—sleep—this means we must begin by setting aside everything we've been told and have come to believe about the nature of sleep. We must then gather factual observations (physical measurements) from the literature and from life, while at the same time omitting any and all interpretations and conclusions. Or as Sargent Joe Friday used to say, "just the facts, ma'am."

Next we'll need to seek out—or design—some scientific experiments, and perhaps do a few of our own. Finally, we'll need to organize this data, looking for "corner pieces"—pairs of complementary opposites.

Is your head breaking yet? If so, I do apologize. I know this is a lot to take in. After all, it's taken me a lifetime to realize and formalize the new method. Please remember this as you read. At the same time, without a better method to guide them, scientists will likely continue to struggle and feel hopeless and lost. As opposed to what it feels like to expect to make discoveries *every time out*. Imagine how that would feel.

Okay. Enough theory. Let's start exploring.

[1] As quoted in Einstein: The Man and His Achievement (1973) by G. J. Whitrow, p. 42

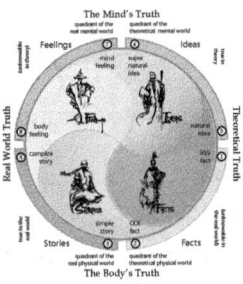

Section 2 - Chapter 10

Sleep
(what can we learn about . . . ?)

What Can Constellated Science Tell Us About Sleep?

For a long time, it seemed inevitable—the older I got, the poorer I slept. Waking five times a night was common. Feeling unrested the next day was the norm. Moreover, had I accepted the commonly dispensed medical advice—that this is simply the way it is; that the older we get, the worse we sleep—then I'd still be struggling through the night and complaining about my sleep.

Is sleeping poorly inevitable? At some point, I began to ask myself. Several of my clients had complained about their sleep. I wanted to be able to help. I then turned to constellated science's, Cartesian Process to open my mind. And the more I allowed for the possibility that science may have missed things, the more curious I felt.

Next, I began to collect information about what science had learned about sleep. At one point, I even bought several huge compendiums of science's current findings on sleep and sleep disorders. I also started to query my clients as to the nature of their sleep. And at some point, I experimented with my own sleep—making small changes, then observing the results.

Much of what I found I did not expect.

Throughout this process, I tried my best to set aside any and all conclusions. Admittedly, with sleep, doing this can be hard—it seems, everyone has ideas about sleep. Over time though, slowly but surely, patterns began to emerge. To my delight, many of these patterns came from seemingly unrelated sources—a sure sign I was on the right track.

Then at some point, I began to summarize my findings in small, exploratory drawings. As I did, new questions began to arise and I grew more excited. For instance, I began to wonder if neurotransmitters came in pairs, and whether alcoholic blackouts might somehow be related to sleep walking. I also wondered how the color and intensity of light affects our two physiological brains—the brain in the head, and the brain in the gut (see Book I: Chapter 4). Could altering the color and intensity of bedroom light improve a person's sleep?

As my list of questions grew, it sparked a genuine desire in me to learn about sleep. I asked myself things like—how does sleep affect your sense of time? Do dreams resemble the early childhood mind? How do paired neurotransmitters promote or disturb sleep, dreams, and moods? And can sleep walking, bed wetting, night terrors, and alcoholic blackouts be different manifestations of the same physiological experience? These questions told me the Cartesian process was working—and that my mind was functioning with the openness of a young child. I literally felt like I couldn't get enough. I'd been captured by the mystery of sleep.

This, then, is what we'll be looking at for the first chapter of section 2. We'll briefly explore the nature of sleep, as I begin to teach you the Cartesian process. Know we'll limit our focus to what happens during this process—this is not a treatise on sleep. Nor is it meant to offer answers. After all, constellated science is the science of discovering questions.

Along the way, we'll create a few "working drawings"—small snippets of logical geometry which organize our findings. We'll also use a rather strange variation of the prototype map, one onto which I've overlaid a figure-eight shape. Some of you may even recognise this shape from mathematics. But if you haven't, know we'll talk more about it in a moment.

Know we'll repurpose this variation of the map many times over the course of this book. Visually, it represents the order of many of constellated science's four-stepped processes. For instance, in later chapters, we'll look at how it reveals the best ways to teach and learn—as well as best practices for doing ostensibly nebulous things like psychotherapy and late-night talks. Finally, as we begin this chapter, please keep mind—we will NOT be doing conventional science. So while I will refer to facts and concepts, we won't be using a hypothesis. Indeed, in most cases, I won't even offer sources.

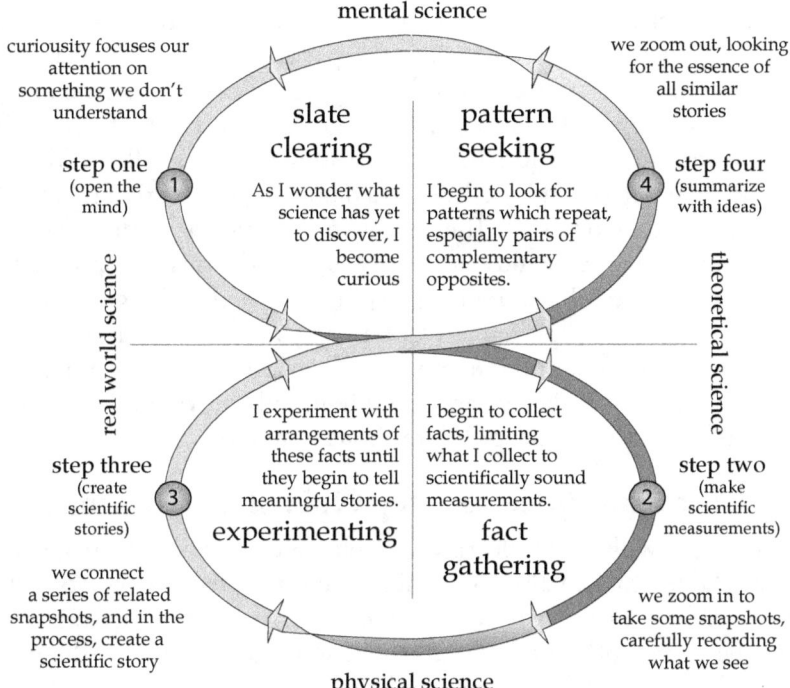

So how can I call this science? My answer is simple. Closed minds ruin science—and hypotheses close minds. Moreover, the entire point of the Cartesian process is to enable scientists to formulate questions. Or as Yoda might say, "Bad questions, there are none. Closed minds, only."

Said in other words, no one can practice—or learn about—science with a closed mind. Or as 19th century British philosopher, Herbert Spencer, (or some say, William Paley) once said ...

There is a principle which is a bar against all information, which is proof against all arguments, and which cannot fail to keep a man in everlasting ignorance—that principle is contempt prior to investigation.

"Contempt prior to investigation" destroys science. Sadly, too many scientific experts have adopted this attitude. This then is the forte of the Cartesian process—it eliminates "contempt prior to investigation."

My Map of the Cartesian Process

When it comes to my diagram of the Cartesian process, admittedly, I have much to explain. For one thing, this map contains neither questions nor answers—only classes of discoveries, and the boundary states which enclose these classes of discoveries. The thing that makes this map unusual though is the way it reveals the relationships between these four classes of discoveries and four states. This and the figure I've overlaid on top of this map—the one which charts the path through the four classes of discoveries.

Now for those for whom this figure is unfamiliar, it's called a Möbius strip. And what makes this oddly-twisted object special is that it has only one surface—only one side. Thus, were you to trace its surface without lifting your finger, you would touch every surface. For this reason, even the smartest of folks can have trouble grasping how such a thing could exist.

For now, please set aside any questions you may have about single-sided objects. Just focus on the numbered steps, beginning with step one: *slate clearing*. What is slate clearing and how will knowing this enable us to scientifically explore sleep? We're about to find out.

Cartesian Process ~ Step One: Slate Clearing

To begin with, slate clearing has but one purpose—to make you curious about a topic. Why? According to constellated science, you must become curious in order to make discoveries. Here, to understand why, recall the pair of opposites I used in Section 1 of this book—the one I used to describe and define open and closed minds.

Open minds are curious. Closed minds are certain.

What does slate clearing have to do with this pair? Quite simply, it clears our minds of all we think we know. And lest this sound hard to do, know it isn't. All it entails is making an "I wish I knew" list. Here, the format for making these lists goes something like this. "If I had access to an all-knowing sleep expert, what questions would I ask?" Now you make your list. Moreover, if you're truly interested in learning how constellated science works, then before reading my list, make your own. Then, when you're done, compare your list to mine.

Here is my slate clearing list. Know it took me a while to make it.

- What does sleep do for me and why do I need it?
- Why do I have two kinds of sleep (REM & non-REM)?
- What does serotonin have to do with sleep?
- Must my sleep deteriorate as I get older?
- How does exposure to sunlight affect my sleep?

- Can too much sunlight impair my sleep?
- Should I take melatonin?
- Can melatonin worsen my sleep? If so, why?
- How do the ratio & levels of CO_2 / Oxygen in my bedroom affect my sleep?
- Does eating before bedtime affect the amount of times I wake up during the night?
- Can the negative effects of long-term shift work on sleep be permanently reversed? If so, what's the best therapy?
- How do the color and intensity of my bedroom light affect my waking up during the night?
- How does bedroom air temperature affect sleep?
- How does the presence and or absence of noise affect sleep?
- Can being drunk or high make it easier to fall asleep?
- Does being drunk or high ruin sleep quality? If so, why?
- Why does worry keep people awake?
- Why do some depressed people sleep too much, and others become unable to sleep?
- Do childhood sleep environments affect adult sleeping habits?
- How do anti-depressants affect sleep?
- What are dreams and why do we have them?
- Why does some sleep refresh us, and other sleep leave us tired?
- How do afternoon naps differ from night-time sleep?
- Why do many people wake up tired after taking sleeping pills?

Admittedly, this is quite a list—far too long for this little chapter. Even so, the length of this list correlates to how much it's cleared my slate.

Cartesian Process ~ Step Two: Fact Gathering

In step 2, we'll begin to explore the questions on our lists. To do this, we'll boil down these questions to a short list of general categories.

Here are the categories I came up with from my list.

- Serotonin, Melatonin, and Neurotransmitters
- Sleep, Light, Sunlight, and Darkness
- Sleep and Air Quality: CO_2, Oxygen, Temperature, Breeze
- REM Sleep, non-REM Sleep, and the Nature of Rest
- What are Dreams and Why Do We Need Them?

One last thing. Constellated science relies largely on discovering geometrically inverse correlations. Even better is finding pairs of geometrically complementary opposites—perfectly inverse correlations. These correlations reveal fractal patterns, and these patterns lead to better questions. These questions then guide the next step—experimenting—which, if we do well enough, can lead to the two questions we'll need to make a map. We'll start with just such a pair of opposites, my first category—serotonin and melatonin. What can we discover about this mysterious pair of chemical messengers?

Serotonin, Melatonin, & Neurotransmitters

The first thing to know about serotonin and melatonin is how they're made. The body makes serotonin from tryptophan—and it makes melatonin from serotonin. This means, whenever the body makes melatonin, melatonin levels increase—while serotonin levels decrease. In effect, the converted Tryptophan either exists as serotonin or melatonin.

This makes serotonin and melatonin physiological opposites—complementary opposites, actually. They come from the same place. But making one destroys the other. Research also shows serotonin and melatonin function as psychological opposites as well. And to see why I say this, consider the class of antidepressants known as SSRIs (Selective Serotonin Reuptake Inhibitors).

SSRIs are used to treat depression—a psychological (and physical) condition wherein people's feelings get compressed into a small space. These folks literally feel too little change when it comes to reacting to life. SSRIs interfere with the body's reabsorption of serotonin, and this fools the body into believing it has more serotonin than it does. And because serotonin "excites things," doing this sometimes helps with depression. People simply feel more over all, and more feeling means less depression.

Melatonin, on the other hand, has the opposite effect of serotonin. melatonin "inhibits things." No surprise people take melatonin to improve their sleep. In some people, doing this calms them down. Interestingly enough, like serotonin and SSRIs, there are classes of drugs which interfere with the way the body uses melatonin as well. They're called, "melatonin receptor agonists." And one of the things doctors use these drugs for is to treat sleep disorders in blind people.

Okay. I admit it. I just gave you way too much information. So before we dig into what all this has to do with sleep—and for those for whom SSRIs are a mystery—please allow me to offer a lay person's explanation for how they work. Keep in mind, I said, "lay person's explanation." Thus, I'll try to follow Nobel prize winning physicist, Ernest Rutherford's, advice—to say things in a way that a barmaid could understand.

An Analogy for How SSRIs Treat Depression

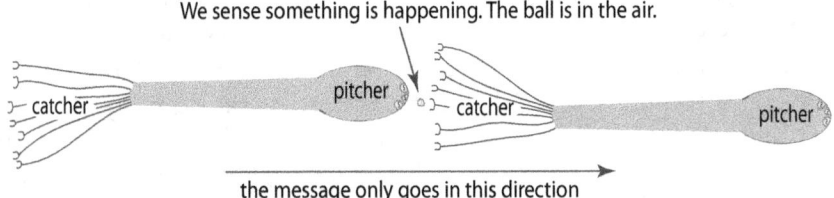

How SSRIs Treat Depression: an Analogy

Nerve cells function like phone lines. They carry messages, each cell in one direction. And when it comes to brain cells, scientists tell us they do this in two ways, electrically—within cells, and chemically—between cells. The word *neurotransmitter* refers to the chemicals which transmit messages between nerve cells. And serotonin is one of these chemical messengers.

Now to see how this works, imagine you're at a baseball game. Currently the ball is in the pitcher's glove, and the catchers glove is open. Now look at the drawing of the two nerve cells I've placed above. Can you see how each cell contains both a pitcher and a catcher?

Real neurons function quite similarly to this baseball game analogy. At the pitcher's end of a cell, there are little "gloves" which hold, throw, and catch-back serotonin. At the catcher's end of a cell, then, there are little catcher's gloves which catch, hold, and return the serotonin to the pitcher end of the previous a cell.

The main thing to know here is what happens when a pitcher in one cell throws serotonin to the catcher in the next cell. A message gets forwarded from the first cell to the second. Once received, the second cell then sends an internal electrical signal from its catcher to its pitcher. In this way, each cell forwards the message on to the next cell. And so on and so forth, down the line, until the message gets delivered.

Why the baseball analogy? Because of what the two cells do next. The catcher end of the second cell returns the ball to the pitcher end of the first cell. Thus, just like a baseball game, wherein the ball gets used again and again, the body reuses the serotonin, over and over again.

This second part of the message-sending process is called "reuptake." During reuptake, the first cell reabsorbs the serotonin it previously sent to the second cell. How SSRIs work is that they interfere with reuptake, thus slowing down this part of the process. This causes the returning serotonin to linger in the space in between cells.

Why does this matter? Because we're only alert to possible messages when the serotonin is in that space. This is similar to how we feel when the ball is in the air at the ball game. We feel anticipatory excitement. So while SSRIs do not increase our actual serotonin levels, they increase the amount of time that we are aware of the serotonin we do have. At the same time, they interfere with our ability to sense time quickly.

The point, of course, is that this is how SSRI's help people with depression. By inhibiting reuptake, they fool the body into thinking it has more serotonin than it does. Moreover, in general, the more people sense serotonin in these spaces, the more alert they become. And hopefully, the more alert they become, the less depressed they feel.

Setting aside any flaws in this logic, can you see how sensing more serotonin might interfere with sleep? Serotonin excites things, remember? And this alertness can interfere with sleep. No surprise, a common side effect of taking an SSRI is insomnia, at least while people get used to taking it. In fact, I once took an SSRI for the symptoms of a bad break-up. After not being able to sleep for four days, I had to go off it.

The insomnia hurt more than the break-up.

Melatonin: a Brief Look

Now let's take a brief look at melatonin, starting with the idea that when I think of the word, I think of the word, "mellow." In a way then, melatonin is the night messenger, while serotonin is the day messenger. Melatonin tells us we're off duty. Serotonin tells us we're on duty.

As for what my research turned up, the first thing is how scientists can't agree on how to classify melatonin. Some say it's a hormone. Some call it a neurotransmitter. And some call it neurotransmitter-like.

Why do scientists often struggle with this kind of confusion? Because nowhere does the present scientific method state how to define terms. Science literally wings it when it comes to most of the words it uses. And most times, they embrace whatever sounds more logical and precise.

In constellated science, terms never get defined by logic alone. They get defined by their geographically logical relationships to all other terms. In part, this is why I keep talking about these pairs of complementary opposites. Without them, we cannot scientifically define our terms.

As for how neurotransmitters work, realize, each cell has many kinds of chemical messengers. Regardless of what messages they send though, they all pass messages the same way. All cells have pitchers at the front end of the cell—and catchers at the back. And these chemicals send messages between cells, while the cells send electrical messages within cells.

Sleep, Light, Sunlight, and Darkness

What determines whether the body makes serotonin or melatonin? It tuns out, there is a tiny, binary switch buried in the center of our brains. This switch is called the pineal gland, and it's hooked up to the optic nerve through the suprachiasmatic nucleus (SCN). The SCN responds to changes in the outside environment, such as the sun going down and the external environment becoming darker. And when the color and intensity of the light we see darkens, this switch tells our bodies to turn serotonin into melatonin. At least until it brightens again, at which point this switch tells our bodies to keep the serotonin it makes.

In this way, the color and intensity of the light we see connects us to our world. In effect, we live in a world which cycles between daylight and darkness—or more accurately, between sunlight and the lack of sunlight. In addition, in between daylight and darkness—at sunrise and sunset—it's normal to see an orangy, transitional light. While during the day, we see light as colorless, when in actuality, it's blue.

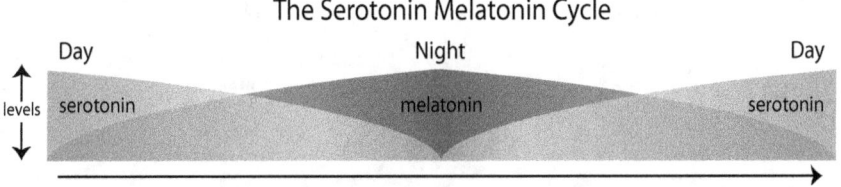

Daylight is Blue Light

Did this last idea surprise you? When I was learning about color, it surprised me too. How can daylight be blue if light is colorless to us? Please know, when these questions arise, it's more proof that the Cartesian process is working. Sadly, we haven't the time to investigate this one here.

What's important to know, though, is that there are meters which measure the color of light. These meters have a scale which equates the color of light to the color of a heated metal. Here, a theoretical metal in a theoretically dark box gets heated to different temperatures. And just like stove tops, as the metal gets hotter, its color changes—from black (cold) to orangy to blue to white (hot).

Orangy sunrise skies equate to something like the color of this metal heated to 2800 degrees kelvin. Old fashioned incandescent light bulbs are close to this color or just a bit higher. Whereas midday skies on cloudless days roughly equate to the color of this metal heated to 5600 degrees kelvin. And on cloudy days, the color of skies can reach 10,000 degrees kelvin.

The point? Our bodies react to these changes in color temperature. And one of the main reactions occurs in our serotonin / melatonin balance. Indeed, one of the best ways to prove to yourself that these two chemicals affect your mood involves adjusting the color of the light that reaches your eyes. Indeed, all you need do to see this in action is to buy two pair of cheap sunglasses—one with orangy lenses, and one with blue-black lenses.

What do these sunglasses do? They selectively block certain color light from reaching your eyes. In this case, the orangey lenses will block blue light, whereas the blue-black lenses will block orangy light. From this, it's easy to create an experiment which roughly approximates the way our seeing the change from sunlight to darkness and back affects us.

In the next step of the Cartesian process: experimenting, I'll describe what happened when I did this experiment. For now, just notice how I'm once again relying on a pair of opposites.

Photosynthesis / Aerobic Respiration

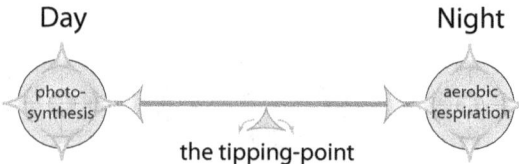

Day: $6\ CO_2 + 6\ H_2O \longrightarrow C_6H_{12}O_6 + 6\ O_2$
Night: $C_6H_{12}O_6 + 6\ O_2 \longrightarrow 6\ CO_2 + 6\ H_2O$

Sleep & Air Quality: CO_2, Oxygen, Temperature, Breeze

Obviously, we are not the only living things that react to changes in the color of natural light. Green plants react to these changes in color as well. During the bluish-light of day, green plants mostly absorb CO_2 and give off Oxygen. And during the absence of light at night, green plants mostly absorb Oxygen and give off CO_2.

Here we see yet another pair of nature's complementary opposites—photosynthesis and aerobic respiration. And for those for whom chemistry is interesting, check out the chemical symmetry. The first reaction is the chemistry of green plant photosynthesis (day). And the second is the chemistry of green plant, aerobic respiration (night).

How does this change in the air we breathe affect our sleep? Let's gather some data.

O_2 & CO_2 (oxygen & carbon dioxide): Measuring Air

A few years back, I bought a wireless scale. This scale measures and records the usual suspects—weight, pulse rate, % body fat, and so on. It also measures what it calls, "air quality"; the amount of CO_2 in the air. To be honest, I'd never thought about this number, let alone how it might be affecting my sleep. After getting this scale, though—and after reading what the manufacturer had to say about this number's affect on sleep—I began to follow this data.

I also began to look for data to substantiate this company's claim. One thing I came across was that in 2012, researchers at the Department of Energy's Lawrence Berkeley National Laboratory found that indoor concentrations of carbon dioxide (CO_2) as low as 600 ppm (parts per million) can significantly impair people's decision-making performance. Evidently, this outcome surprised many people, as previously, it was thought this number had to be much higher before it negatively affected us.

What's my home like, CO_2 wise?

According to my scale (and three years of data), on a really good day, this number varies from just under 400 to just under 600. Here, we're talking a summer day, with several windows wide open all day long. On the other hand, one day a few months back, it varied from 302 to 1258. My gas drier had malfunctioned, driving the number into the red.

Based on the Berkeley Labs study then, the air in my home that day was well over the point at which it would negatively affect my brain. Did it? I'm not sure. I was too worried my house would blow up to notice anything as subtle as my decision-making acumen.

Thank goodness it happened during the day.

Of course, this only led to more questions like, how exactly do we react to the changing levels of these two gasses in our air? For instance, if raising the CO_2 levels negatively affects us, does raising the O_2 levels positively affect us? What about breathing too much O_2?

What About Sleep and Breathing?

The main function of breathing is exchanging carbon dioxide for oxygen. Scientists call this process, "ventilation." Not surprisingly, I'm told all breathing-related reflexes are triggered by the CO_2 concentration in our blood. I've also read a study wherein briefly increasing the CO_2 levels in the air being given to pre-term infants caused the opposite reaction—some of the infants briefly experienced apnea—a pause in their breathing. What this means is far beyond me, other than to notice yet another pair of opposites.

Of course, the thing to take away here is that we've found yet more unanswered questions. The main question is, how are sleep apnea and CO_2 levels correlated? If blood levels of CO_2 trigger breathing, and if air levels of $CO2$ trigger pauses in breathing, then what does the CO_2 actually interact with, and is there a level of sleep past 4—a level near death?

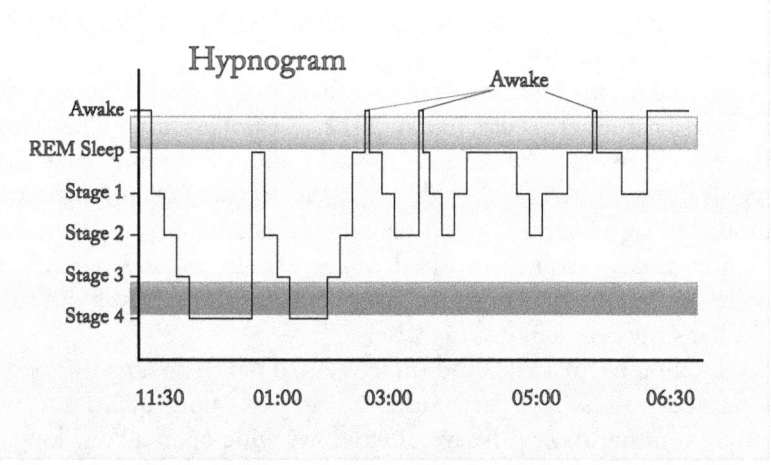

REM Sleep, nonREM Sleep, and Rest

Speaking of sleep "level 4," most people know we go through changes during our sleep. Indeed, most people have heard of REM sleep—and most people know that REM stands for "rapid eye movement." They also know we cycle between REM sleep (the kind the hypnogram pictures with no levels) and nonREM sleep (the kind with 3 or 4 levels). Other than these few things though, few people can point to what makes these two kinds of sleep different—let alone that they are literally yet another of nature's complementary opposites.

Then there's the thing about which kind of sleep we start the night off with. Newborns and narcoleptics begin sleeping with REM sleep. The rest of us begin sleeping with stage 1, nonREM sleep. Of course, some researchers argue that in the case of babies, this is not actually REM sleep. They call it, "active" sleep. Know we'll talk more about this idea in a bit, as it's somewhat important to see how they came up with it.

Now look at the "hypnogram" drawing. In it, you'll see how sleep scientists chart these changes during sleep. On the left, the diagram begins with awake, then descends through nonREM stages 1 through 4. After a stay at level four, it then shoots straight up to REM sleep. Then after a brief stay in REM sleep, it descends through the nonREM levels again, only to rise up into REM once more.

The first thing to see is that nonREM and REM sleep cannot coexist or overlap. Either people are in nonREM sleep—or they are in REM sleep, making them two separate states. This oppositeness is why I referred to these two kinds of sleep as yet another of nature's complementary opposites. Thus like serotonin and melatonin, we can learn much about nonREM and REM sleep by exploring what makes each sleep state different.

Let's start with time.

Over the course of the night, nonREM sleep periods get shorter while REM sleep periods get longer. In essence, the sleep we get early in the night is mostly nonREM, and the sleep we get toward the end of the night is mostly REM. This raises the question—if we get too little sleep, does the sleep deficit lie mainly in REM sleep? If so, then does this account for the times when I've taken an afternoon nap and gone straight into REM sleep? Was my body making up for what I lacked when I was in bed for too short a time? Or are naps a time wherein people revert back to an early babyhood state?

The biggest question of all, of course, is why we need to sleep? What does sleep do for us—and why do we need two kinds? As a side note, I read that during REM sleep, the eyes of congenitally blind people move the same way our eyes do. Which makes sense if we humans need this kind of sleep. On the other hand, sleep scientists claim our eyes are following what we dream during REM sleep. So is this proof congenitally blind people see things? If so, what are they seeing? Or do eyes moving during REM sleep mean something else, something we sighted people don't normally recognize? And are babies and blind folks doing the same thing?

Then there's the idea I mentioned previously—that a common side effect of SSRIs is that they inhibit REM sleep. Why? This infers an inverse correlation between increased levels of serotonin and decreased REM sleep. So does this mean that REM sleep correlates to a significant, temporary decrease in serotonin levels in us? Does this also mean that if the body *produces* more melatonin, that the person will get more REM sleep, all other factors being equal? And how does decreased REM sleep affect depressed people?

Are you beginning to see how the Cartesian Process opens minds? Can you also see how it does this in ways the current method can't dream of doing, no pun intended. To wit, I have yet to find professional research which raises, let alone addresses these questions. And I, for one, would love to know what all these things mean about sleep.

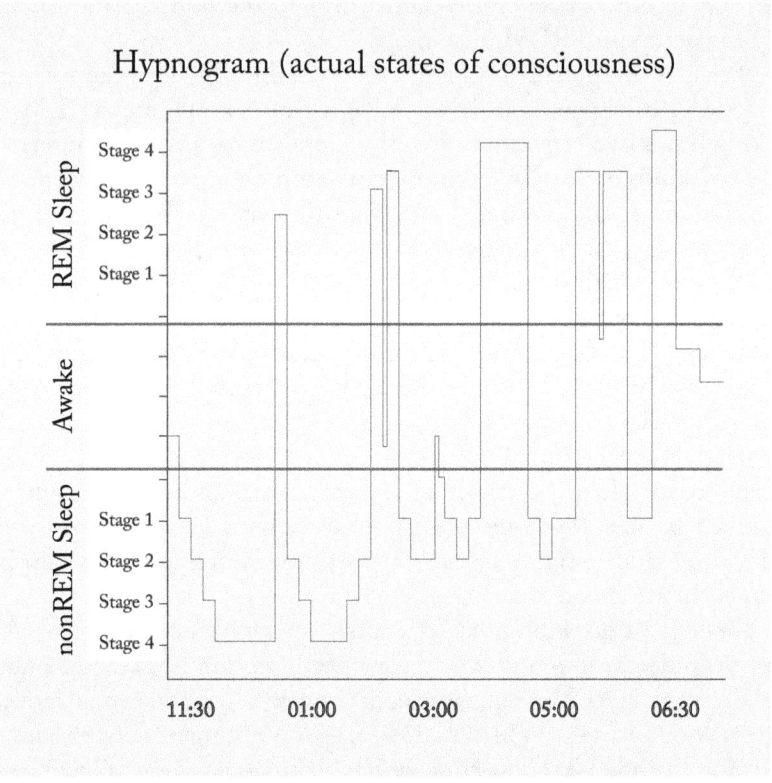

Is This a Better Way to Order Hypnogram Levels?

Lastly, did you notice how the first hypnogram charted REM sleep just below awake? During the writing of this section, I read in several places that the positioning used in hypnograms is an arbitrary choice. In part, this choice seems to come from the idea that REM sleep is similar to being awake. And in part, it seems to be because we usually wake up right before or after REM sleep.

Here's the thing. Given that REM sleep and nonREM sleep are complementary opposites, is it possible REM sleep should be placed on the other side of awake? In other words, does the state of consciousness we call "awake" actually lie in between the states we call REM sleep and nonREM sleep?

Now before you write this idea off as ridiculous, take a look at the snippet of logical geometry I've placed above, the one labeled, "actual states of consciousness." Know we'll explore this possibility in depth in a future chapter when we look at the mind and consciousness.

How About Sleep & Dreams?

At this point, I hope you're beginning to see what a big part the third geometry plays in discovering relevant questions. By this, I mean, I constantly look for nature's pairs of opposites. In nature, it's unusual to find symmetry of any kind. Thus when you find it, you know you've located an important relationship—be it serotonin & melatonin, REM & nonREM sleep, Oxygen & Carbon Dioxide, or photosynthesis & aerobic respiration.

What about dream states, then? Are there opposite dream states as well? Knowing logical geometry, my first reaction is to say, yes, there must be. Here, the first question to come to mind is whether day dreams are the opposite of night dreams? Then again, what about hypnosis? Is hypnosis like day dreams? Is it also an awake dream state? Or can hypnosis be an excursion into one of the levels of REM sleep? For that matter, can we even know whether we are awake or asleep during hypnosis?

What about hallucinations? Are hallucinations "awake" dream states too? Or are they sudden-onset periods of REM sleep with no memory of where or when we've exited the awake state? What about spiritual visions? Near death experiences? Are these things dream states too? And what about babies just before they're born? Are they in a dream state only to awake when they're born?

Why include the dream-state-like things we do while we're awake? After all, aren't we exploring sleep? Well if being asleep and awake are a pair of natural opposites, then we must explore both kinds of dreams. Indeed, only by identifying all the puzzle pieces—and how they fit together—can we begin to know what dreams are. In truth, it may just be that the alternate hypnogram will give us a way to see these very relationships.

Finally, why is it most dreams occur during REM sleep (80%) and not during nonREM sleep (20%). Know I say "most" because the dreams we call night terrors occur only during nonREM sleep. And what about when people sleep walk? Is this some strange kind of nonverbal dream state?

A Look at Babies' Sleep States

Before moving on to the third step of the Cartesian Process—experimenting—I'd like to briefly address how babies' sleep states differ from ours. To begin with, babies' sleep cycles are shorter. Ours are about 90 minutes. Theirs are about 50-60 minutes, at least for the first nine months or so. Moreover, while sleep scientists say we have 4 or 5 levels of sleep, babies are said to have only two—active and quiet. Here, babies begin with active sleep. Their eyelids flutter. They take somewhat rapid, irregular breaths. They occasionally move. They briefly grunt, moan, and cry. And they also awaken more easily.

Sounds like REM sleep, doesn't it?

About half way through a sleep cycle, babies then enter the second state: quiet sleep. Here, babies' breathing slows down and is more even. They move very little. And there is no eyelid fluttering. Also, they are less likely to be awakened by noise and other disturbances.

Quiet sleep represents the end of the baby's sleep cycle. When it's over, babies either wake up or begin the cycle again, by re-entering active sleep. As babies mature, their quiet sleep time begins to differentiate into distinct nonREM stages. Also, their sleep cycles lengthen, and they spend less and less time in active sleep.

Know this change happens slowly. It takes several years for children's sleep cycles to look like those of adults. So although nonREM sleep stages emerge by 6 months, sleep cycles don't reach the normal 90 minute mark until children are ready to enter school. From then on—and throughout early childhood, kids spend a lot of time in REM sleep. So while adults only spend about 20% of the time in REM sleep, newborns spend about 50% of the time in REM sleep, and three year olds about 30%.

One last thing. I mentioned the idea that some—not all—scientists claim the active sleep of babies is not normal REM sleep. So although babies show all the same behaviors—including rapid eye movements—these scientists say this cannot be REM sleep. Why? Because they say babies aren't mature enough for their eyes to follow their dreams. Obviously, something is different. But where's the evidence? And yes. Babies have yet to develop most normal motor skills, including depth perception. But people born blind never develop normal visual motor skills either. So is the active sleep of congenitally blind people not actually REM sleep either? All I can say is, to me, conjecture based on logic alone is at the very least, poor science.

Cartesian Process ~ Step Three: Experimenting

At this point, I hope you'll forgive me. I've definitely overdosed you on ideas. Moreover, despite this step being called "fact gathering," admittedly, I've omitted most facts—and used ideas to summarize groups of facts. Know that ordinarily, taking this step would involve listing all these facts. But in a book which primarily focuses on teaching a method, doing that would be impossible. This book would be over 2,000 pages.

Step three—experimenting—on the other hand, must result in facts. And in case you have yet to read Book I, I define facts in the following manner. To be a fact, a thing must reference three things: a specific time, a specific place, and a specific observable event. Miss any of these three things and what you're referring to is not a fact. It may be an idea and it may be well grounded in science. But even a well grounded idea is still not a fact. At the same time, to create an experiment, you must be able to describe this experiment with ideas. It still must result in a fact or facts though.

Here's the first example. I'll begin with ideas, then offer a fact or facts.

The Serotonin / Melatonin Sunglass Experiment

Most people know the colors around us affect our moods. Here, it's not hard to find scientific research to support this claim. Psychologists have been observing—and trying to make sense of—this effect for years. Even so, the third step of the Cartesian Process requires me to do my own experiments as well. This experiment was my first in and around sleep.

As usual, I began with a question. I asked myself if I could use colored sunglasses to alter people's serotonin / melatonin balance. After all, this change is one of the main components in sleep and wake states. I also wanted to know if people could notice this shift when it happened. So knowing our serotonin / melatonin balance is largely keyed to the color and intensity of light, I drew on something I learned in 7th grade science class—Mr. Branch's basics of color theory. And for those for whom these basics are long forgotten, here's the five cent tour.

The Basics of Color Theory

First, like all natural things, *color* is based on a pair of complementary opposites. The opposites I'm referring to here are the color of light—and the color of pigment (paint). With light, color is emitted—and every color is transparent and made of light. But with pigment, color is absorbed and reflected—and every color is opaque and derives from what is left.

This means, with light, when we add colors, we add light. So things get brighter. But with pigment, when we add colors, we add pigment. So things get darker. With light then, the colors we see are the sum total of all the colored lights combined. But with paint, the colors we see are the sum total of whatever hasn't been painted over, in essence, what hasn't been subtracted out. Thus with light, colors add—and adding all the colors together results in the color white. Whereas with pigments, colors subtract—and adding all the colors together results in black.

Said more simply, with light, *adding colors adds color*. And with paint, *adding colors subtracts color*. And of course, all this sense of color changing occurs in the mind, not in the world. In the world, only wavelengths change.

Cheap Sunglasses

Skipping that last little nugget about the mind, can you guess which part of color theory I needed to use to explore sleep? In truth, I needed to combine all of it with my knowledge of serotonin, melatonin, and sleep. My goal was to manipulate the light reaching my eyes. By doing so, I hoped to simulate the changing colors we experience in a day.

At this point, I knew I'd need to get something which I could see through—but which also added a tint which shifted the colors of what I was seeing. I also needed to cause two kinds of shifts, one towards the colors I might see at sunrise / sunset, and one towards the colors I'd see at midday. Here the choice was obvious. I decided on cheap sunglasses.

The first half dozen I bought were called, "blue-blockers." These glasses have lenses covered in an orange tint. Moreover, since this tint was pigment, these glasses subtracted blue light and made the light I saw less intense. They also made everything look sort of orangy, something like campfire light or the sky at sunrise or sunset.

I also needed glasses which had the opposite effect. So I bought half a dozen sunglasses with bluish-black lenses. I hoped these glasses would function like "orange blockers," in effect, decreasing the intensity of the light and giving everything a bluish tint. Why? Because daylight is blue, remember? And because transitional light (sunrise and sunset) is orange. And since orange tinted lenses block blue light, they cause us to feel as if we're experiencing transitional light. Whereas blue tinted glasses, which block orange light, simulate the color of the light we see at midday.

Can you see the complementary opposites in play at this point?

For one thing, the two kinds of sunglasses acted as external, complementary opposites. Literally, each type of sunglass subtracted color at the opposite end of the visible light spectrum. And no, this oppositeness was not the science-lab variety, the perfectly linear kind. But because tipping points allow us to accurately measure real world, complementary opposites, this didn't matter. No statistics necessary, and I could be 100% sure of the measured outcomes.

A second pair of opposites in play were the neurotransmitters involved. Here I hoped the sunglasses would affect a physiological pair of internal, complementary opposites—our serotonin / melatonin balance. Ultimately, I hoped this would also affect a pair of psychological complementary opposites—our alert and mellow moods. More important, I hoped that people would be able to notice these changes.

At this point, I was ready to design the actual experiment. Know that how I went about this points to yet more differences between conventional science and constellated science. In truth, designing experiments is one of the trickier parts of doing any kind of science. And the hardest part of all is designing experiments which yield results with a high degree of certainty, but which, at the same time, translate to the real world.

Conventional science does the first part fairly well. Really well, in fact. But they struggle achieving the second criterion—arriving at results which translate well to the real world. And yes, sometimes their results do appear to translate well to large groups. But to individuals?

Enter constellated science.

The Problem of Confounding Influences

Pairs of opposites. Pairs of opposites. More gosh-darned pairs of opposites. Why? Because over and over, when we contrast and compare pairs of opposites, we make discoveries. Moreover, before I explain the experiment, I need to first account for yet one more pair of opposites. This time, though, it's not an element of what we're experimenting on. It's an element of the experiment's design. Here, I'm talking about the ways in which the two scientific methods deal with what conventional scientists call, "confounding influences."

What are confounding influences?

Conventional science sees confounding influences as anything which might obscure or confuse an experiment's outcome. Or more specifically, anything which might prevent them from arriving at *answers* with a high statistical probability for being correct. Conventional scientists focus on answers, remember? And they rely on statistics for these answers.

The thing is, confounding influences—meaning, everything these scientists can't or won't account for in the real world—ruin statistics. Indeed, scientists admit they struggle to statistically account for more than two variables. In the real world though, things often have an infinite number of influences, many of which cannot reliably be accounted for, let alone measured. This forces scientists to design experiments which isolate the thing they're exploring from the influences of the world. Here a good way to understand why doing this causes problems is to consider the old story about the drunk who has lost his keys. Do you know it?

A drunk has lost his car keys and is searching for them under a street light. His friend comes along and asks him what he's doing. The drunk answers, "I'm looking for my keys." The friend then asks the drunk where he last saw his keys. The drunk answers, "across the street, over by my car." The friend then asks, "then why are you looking for them under the street light." And the drunk answers, "because that's where I can see."

In effect, conventional science only designs experiments under street lights. By doing this, they hope to eliminate any and all things they can't account for. Admittedly, by doing this, they often arrive at repeatable, reliable outcomes. And to be honest, this design does have its place in science. It can be quite revealing to see how a thing behaves outside of its own environment. However, when it comes to learning the nature of living things, isolating them from their environment guarantees unnatural and often misleading results. Things simply do not behave the same way in unnatural settings. Think bears in a zoo compared to those raised in the wild. Indeed, no living thing lives in isolation from it's environment other than when people cause this. Moreover, the nature of all living things IS the sum of their influences on other things, living or otherwise—plus the influences other things, living or otherwise, have on their nature.

From this definition it's easy to see why laboratory experiments have a limited ability to predict the nature of real world things. Laboratory experiments lack nature's symmetry and infinite connections.

A Solution for Confounding Influences

Constellated science sees experiments done in isolation as last resorts. Indeed, what conventional science sees as confounding influences, constellated science sees as important and necessary. Why? Because we're talking about "natural" science, not artificial science. So isolating a thing from its natural environment changes its nature.

Conventional scientists might argue, "but nothing in the real world repeats the same way twice." And this is true. This is why, to them, when it comes to living things, the lack of linear results equals unreliable conclusions. The problem of course is that conventional scientists focus on answers, not on knowledge. To them, answers are their holy grail. Natural knowledge—when possible—is but a fortunate side effect.

What's the solution? Build your experimental designs around complementary opposites. Sadly, conventional scientists have yet to see how doing this solves the measurement problem. If your experimental designs employ natural complementary opposites, then your outcomes will always yield results which parallel a landed coin in a heads / tails coin toss. By this, I mean, completed experiments will always lead to observed outcomes which are 100% clear and certain, even in the real world. This then solves the problems inherent in isolation labs.

In this way, constellated science's experimental designs avoid many of conventional science's problems, including those associated with measuring things in the midst of an infinite number of variables.

Finally, the First Sleep Experiments

I doubt you'll be surprised at where I began my experiment. I decided I needed to try it out on myself, so I could know how to conduct the experiment with others. Not that I'm the perfect subject. Far from it. But it helps to gain a baseline of your own experiences before testing others.

I remember it being bright and sunny in my office that day. I began by trying to notice any tension in my face and eyes, with no glasses on. To be honest, at first, I didn't notice much of anything.

Then I put on the blue-blockers, and the world turned orangy.

I then sat with these glasses on for the next few minutes. I then told myself to focus on my reaction as I took the glasses off. To my surprise, I noticed my eyes and facial muscles tensing up a bit. "Was the light too bright," I wondered. I tried turning away from the window. Then I repeated the experiment. And again, I felt my eyes and facial muscles tense up.

The thing is, when I started, I had not noticed any tension. I became aware of it only after wearing the blue blockers for a few minutes, then taking them off. Could it be the glasses were decreasing the light's intensity, and then taking them off made the light seem too bright? I wanted to rule that out. So I backed away from the window even further. Then I tried the experiment again.

To my surprise, the effect of taking the glasses off was still the same. Indeed, I might have felt it a bit more. Moreover, the more I did this experiment, the more I noticed what happened when I put the glasses on and took them off. My eyes and facial muscles would relax with them on, and tense up when I took them off.

I also noticed that putting them on made my mental mood relax as well. This observation was important, as seeing both my mind and body reacting similarly further validated what I was seeing. Moreover, when I took off the glasses, I felt the opposite mental reaction. I felt my mind tense up.

Did you notice how I did not treat any of what I was finding as answers? As I've said, constellated science is all about questions—answers close the mind. No surprise then that, at this point, even more questions began to arise. I chose to ignore them for now and to continue to gather more data. After all, wasn't this the entire point for doing these experiments—to gather data?

Next I tried the orange blockers and by now, I was getting good at noticing changes. Within seconds of my first time putting them on, I noticed my eyes and facial muscles tense up. Taking them off then had the opposite effect. I felt my eyes and facial muscles relax the minute I took the glasses off.

Again, notice, I am making no claims as to what was causing these things to happen. Indeed, cause and effect speculations have no place in constellated science. Doing this would be like speculating on how each star in the big dipper came to be located where it is located. Who cares? The only thing I needed to notice was how these glasses were affecting me.

Over the course of the next few years, I repeated these experiments with close to one hundred people. A few times, I did this in groups. But mostly, I did it in my office with just one other person. Out of this hundred, two people had trouble noticing any changes. But most were surprised by what they felt. More important, these folks all reported having the same reactions I did. Blue blockers on—less tension, mentally and physically. Blue blockers off—more tension, mentally and physically. Orange blockers on—more tension, mentally and physically. And orange blockers off—less tension, mentally and physically.

Finally, for those for whom such details are important (me, for one), know that I told these folks nothing about what to expect, nor even why I was asking them to do these things.

Thank God I have access to so many open minds.

My Experiments with Light & Waking Up at Night

As I mentioned at the beginning of this chapter, at the time I did these experiments, I was struggling with my sleep. I was waking up eight to ten times a night, and I found it hard to accept that this would be my sleep pattern for the rest of my life. This said, know I hadn't had sleep in mind when I did the sunglasses experiments. I'd simply been curious as to the role light played in mood changes.

Also, I'd previously experimented with how changing the colors of the things in my office affected therapy. In fact, I'd ended up adjusting the color balance in my office so it matched the varying sensitivities of the human eye—mostly middle green with some yellow and a bit of reds and violets. This color balance elicited many unsolicited comments, most of which centered on how much calmer my clients felt just entering the room.

This raised many questions in me. In particular, I wondered how much we might improve our lives were we to match our environments to the natural balances of our five senses. My experiments with the colors in my office seemed to hint at this. It wasn't until I'd done my sunglasses experiments though, that I began asking myself how the color and intensity of light had been affecting my sleep. This time, designing experiments to explore this question required no purchases at all. My normal routine had been to sleep in a pitch-black room. I, in fact, had been sleeping this way since I was ten years old.

As I thought about this, I began to wonder how sleeping in such a dark environment had been affecting my serotonin / melatonin balance. If melatonin is keyed to the absence of light, could my melatonin levels be too high? Was this even possible? And if it was, how would it have been affecting me?

Not coincidentally, at the time, I'd been working with someone who, for years, had had trouble noticing his feelings. And for those who are not familiar with the extreme version of this condition, trouble noticing feelings can be a sign the person has *dysthymia*—a condition wherein this person gets depressed in ways which parallel a functional alcoholic. In effect, my client had been a functioning depressed person for most of his life. And for about a year, he'd finally broken through this depression, and had been feeling an almost normal range of emotions.

I mention this case as I clearly recall this man telling me he'd been having trouble sleeping. So he'd decided to take melatonin. Sadly, within two weeks of taking it, he'd regressed to the point where he was once again depressed and significantly detached from his emotions. At which point, I suggested he speak to his psychiatrist about discontinuing the melatonin to see if this cleared this up. Sure enough, within two weeks, he was back to a normal range of emotion.

Here again, I drew no conclusions from this. I focused only on the questions this raised. But the effect I'd witnessed in this man's range of emotions made me wonder even more whether it was possible to have too much melatonin. With this in mind, I booked a flight to an Andrew Weil sponsored conference titled, The Integrative Mental Health Conference. In part, this was because several of the presenters were experts in sleep medicine.

Let me cut to the chase. Every expert I asked told me the same thing. You cannot take too much melatonin. Indeed, when I tried to refer to my own observations, they quickly and cavalierly dismissed me. Ah, how it must feel to have sure and certain knowledge. Unfortunately, when a mind is certain, this mind is closed—and closed minds do not make discoveries, let alone teach well. Sadly, I went home more confused than ever, but firmly resolved to investigate further. It seemed illogical that there could be no downside to taking too much of anything, let alone a major neurotransmitter / hormone / sort-of-neurotransmitter.

What experiments did I end up doing? First, I need to tell you the series of thoughts which lead me to do these experiments. When I'd seen my client's depression return, I asked myself if my own lifelong, dark and dismal bedroom environment had anything to do with me having dysthymia for much of my life, including for my entire childhood. Could sleeping in a too dark room for too long push a person's melatonin levels too high? Depression is the lack of emotional life. Melatonin had seriously impaired my client's ability to feel emotions. Could melatonin calm down a person's emotions too far? For that matter, what role did melatonin play in my client's depression?

In the midst of asking these questions, I then had the experience which led to my experiments. I noticed I'd experienced good sleep at a time and place in which I didn't expect it. To wit, one afternoon, I decided to take a nap on my couch and had slept the sleep of the dead. I also woke up from this nap more refreshed than I had in years.

Then it hit me. I'd been napping in bright sunlight. And I'd had no trouble going to sleep. Was I simply that overtired? Or had the sunlight had a positive affect on my sleep? If it had, then why was it so hard for me to fall asleep with a light on in my bedroom?

Questions. Questions. Questions. The Cartesian Process at work.

The Hall Light / Bedroom Light Experiments

I began these experiments by leaving my bedroom door open and the hall light on. This light was a single, 60 watt incandescent bulb whose color was about 3,000 degrees kelvin—orangy light. At first, when I tried to fall asleep, I was so disturbed by this orangy light that I worried I'd never fall asleep. Even through my closed eyelids, it seemed too bright.

Somehow, though, I did fall asleep, and that night I slept through the night for the first time in years. Obviously, one night doth not an experiment make. So I kept at it for several months. At first, I continued like this for several days. Then I tried decreasing the intensity of this light, by half closing my bedroom door. At some point, I also tried changing the color of this light. I changed out the tungsten light (orangy light) for a daylight bulb (bluish light)—something like 5,600 degrees kelvin.

Finally, I started to wonder whether the amount of sunlight I got during the day meant I needed to alter the amount of light I got at night. I'd suffered from Seasonal Affective Disorder from before they'd named it this. So I knew the amount of daylight I exposed myself to affected my moods. To account for this, I began to note the degree to which I experienced sunlight during the day. I then tried to see how this correlated to the quality of my sleep.

One thing I noticed throughout this time was how much my responses varied. Some nights I woke only once or twice. Some nights I woke more times. Curiously, after the first night, not once during this time did I sleep through the night. But in the end, I observed several patterns.

To begin with, for me, it seems, adjusting my bedroom light can affect the quality of my sleep. As does the amount of sunlight I expose myself to during the day. I also cannot get over how much getting good sleep changes my moods the next day. I am also surprised at how I'd never realized what all this infers about serotonin / melatonin balances changing. As well as how this balance may have been affected by my having slept in a completely dark room all my life.

Overall, the thing that's stayed with me the most is how my experiences with daytime naps can be so opposite from my night sleep. I was—and am—pretty good at what sleep experts call, "sleep hygiene." I go to bed within fifteen minutes of the same time each night. I almost never eat during the three hours before I go to bed. And I never sleep with a TV on. Or expose myself to intellectually provocative—or troublesome material—right before bed, if I can help it.

In effect then, when it comes to night sleep, I follow most of what sleep experts recommend. Oddly, when it comes to taking naps, I seem to follow none of these night sleep recommendations. My naps occur at all random times. I can eat or read right before I nap. Yet doing these things seems to have no negative affect on the quality of these naps. Why? And why does napping in bright sunlight not keep me awake, let alone make it harder to fall asleep?

Know I've since taken quite a few clients through these same sleep / light experiments. For those that can tolerate this tedious process, most manage to improve their sleep. In particular, one man who'd been a shift worker for ten years—and who had then suffered from poor sleep for ten years afterwards—became able to sleep normally. And yes, this took us close to a year and involved many experiments. Besides working with light, we repositioned his bed, opened and closed blinds, and altered bedtimes. However, like all of constellated science's experiments, these experiments not only helped. They also led to new lines of questioning, some of which only made sense after completing this process's next step. Know we'll talk about some of what I discovered toward the end of this chapter.

The Childhood Sleep-Environment Surveys

Not too long ago, a couple came in and talked about possibly sleeping in separate rooms. The wife brought this up and notably, did so with no blame, anger, or resentment. The husband had been away for several days and she reported having slept noticeably better. And after one day of him being home, she woke up so tired that a friend had commented on how unrested she looked.

I then began to explore with them what their childhood sleep environments had been. She'd slept in quiet, dimly lit bedrooms. He'd slept with the TV on. She'd disliked air conditioners and preferred open windows. He preferred air conditioners. These responses reminded me of the questions I'd asked myself about my own childhood bedroom experiences. And when the three of us compared our experiences, it seemed, many of the same things which disturbed this woman—brightly lit bedrooms, noisy bedrooms, air conditioners on, having a TV on—disturbed me too. While the opposite things disturbed her husband—completely dark bedrooms, quiet bedrooms, not having an air conditioner on, not having a TV on, and so on.

Of course, the question I was left with is, why all the complementary sleep patterns? This led me to ask what I still find is a rather strange question. How often do couples' sleeping habits contain these patterns of opposites? Here, I'm not just referring to the patterns of things which disturb people's sleep. I'm also referring to the patterns of things which people need in order to have good sleep.

Me being me, I then began to look for these same kinds of patterns in other couples. I'd been seeing couples for more than two decades, and sleep incompatibilities are common. Because of this, I've always asked couples about their bedroom temperature ("Do you sleep with blankets in the Summer?"), and light and noise ("Do you sleep with the TV on, or with some other kind of light on?"). And often—not always—couples exhibit the same pairs of opposites—with light, sound, temperature, TV, late-night eating, circadian rhythms (timing of wake / sleep states), and so on.

Indeed, I recently taught a small group (n=9) and asked them to compare themselves to their partners using three of these questions—temperature warm / cool, TV on / off (light & sound), and wake / sleep timing. Of the twenty seven possible opposites these folks could have answered (3 questions times 9 people), they answered twenty six of twenty seven as clearly-tipped opposites.

Over the years, I've asked myself many times how this could happen. Is there something in human nature that leads most of us to be with our opposites? After many years of observing and inquiring, though, I am still no closer to an understanding. Which, in and of itself, is a good thing, really. My curiosity is alive and well. Still, I'd like to uncover some connecting patterns sometime soon. How about it, Universe. A little help, here?

Cartesian Process ~ Step Four: Pattern Seeking

In a way, the final step in constellated science's *first* process functions a lot like a mini version of the entire *fourth* process. In both cases, the goal is to find patterns of connections in data. Here though, you're only looking to create snippets of logical geometry. Whereas in constellated science's fourth process—constellating—your focus is to find parallels between a new map and every other scientific constellation.

Unlike constellated science's fourth process then—which focuses entirely on finding parallels between maps, the Cartesian Process's fourth step looks to find patterns of facts which lead to a new map. Please keep this in mind as we progress though this step then. We're not looking to parallel other maps. We're only searching for patterns within the data we've gathered.

Re-Listing My General Categories

In a moment, I'll begin to list the pairs of opposites I've found. Here, I'll use a few words to describe each of them. Then when I'm done, I'll look to create a few snippets of logical geometry. Hopefully, at some point, these snippets can lead to a map.

For now, my focus will be on creating this list. Know the word to keep in mind will be, "psychophysical." This word refers to the idea that our mental and physical worlds share the same set of universal rules. Can you see how this idea led to logical geometry? Moreover, can you see how logical geometry looks to build bridges between these two worlds—between constellations of logical words and constellations of seminal shapes?

First, let's re-list the five categories I planned to explore. They are:

- Serotonin, Melatonin, and Neurotransmitters
- Sleep, Light, Sunlight, and Darkness
- Sleep and Air Quality: CO_2, Oxygen, Temperature, Breeze
- REM Sleep, non-REM Sleep, and the Nature of Rest
- What are Dreams and Why Do We Need Them?

Next, let's work through the list of complementary opposites we found. Here we'll need to do our best to let our minds free associate. We'll also try to learn something about how these categories connect, starting with the point of it all—to see what we can learn about sleep itself.

Cartesian Process ~ Our Working List of Opposites

I'll begin by listing the pairs of opposites I've mentioned in the text. I'll then explore these opposites, looking for parallels between them. Our questions will then center on finding questions which explore the nature of these parallels. Remember though, we're NOT looking for answers. Only questions.

Here's what I've found so far.

1. States of human consciousness (asleep / awake)
2. Circadian states of earth environment (day / night)
3. Sunlight / Darkness (high intensity light / low intensity light)
4. Daylight / Transitional Light (bluish light / orangy light)
5. Serotonin / Melatonin (day / night) levels (melatonin is made from serotonin)
6. Serotonin promotes wake / Melatonin promotes sleep
7. Photosynthesis / Aerobic Respiration (Air O_2 / CO_2 balance)
8. Human Respiration / Green Plant Respiration (O_2 / CO_2 levels)
9. Human Respiration (day / night <> blood O_2 / CO_2 levels)
10. REM (mind on/ body off) / nonREM (mind off/ body on)
11. Sleep Start State <> babies (REM) / adults (nonREM)
12. Sleep Patterns / Couples (room temp, timing, noise, light, TV)

Now I'll list a few of the complementary opposites I found, but did not have space to mention in the text. Again, remember, we are NOT looking for answers—only questions.

13. Blood CO_2 up > PH more acidic / CO_2 down > PH more alkaline
14. Body Temp before / after Tmin (lowest body temp per 24 hours)
15. Eat close to bed / fast before bed (weight gain / loss are inverse)
16. Memory consolidation (body memory / mind memory)
17. Melatonin / Cortisol levels (daylight / darkness are inverse)
18. Acetylcholine / serotonin levels (REM / nonREM are inverse)
19. Glutamate (excites CNS) / GABA (inhibits CNS) (GABA made from glutamate)
20. Norepinephrine levels (REM / nonREM are inverse)
21. Dream character & content (REM / nonREM are inverse)
22. Heart rate & blood pressure (REM / nonREM are inverse)
23. Respiration: pattern & stability (REM / nonREM are inverse)
24. Body temperature (REM / nonREM are inverse)

Cartesian Process ~ Outcome: Sleep Questions

Finally, we arrive at the outcome of the Cartesian Process—our working list of opposites, and a preliminary list of questions. The opposites are the raw materials for a new map, and the questions will insure our minds remain open during the next process. In a way then, these questions take the place of conventional science's hypotheses. Here, conventional science focuses from start to finish on discovering answers, beginning with a guessed answer (the hypothesis) which is then explored to see if it is a true answer. Whereas constellated science focuses start to finish on discovering questions, beginning with the prototype question (what can we learn about . . . ?), which is then explored to discover patterns of connections hidden in these questions.

At this point though, all we'll focus on is organizing what we've found so far. To do this, we'll use a brief narrative of the entire first process. Here, we'll break down this narrative into the same five sections we just employed, starting with the role of neurotransmitters in sleep. But first, I need to remind you of a few things.

One thing concerns the data I've mentioned. To wit, in every chapter in this section—and in every chapter in the entire book, actually—I refer to summarized data. Know in no way am I claiming this data is true.

Why refer to this data then? I needed to use it to pose examples of the method. Thus none of what's in this book is intended to be seen as—or lead to—answers. What this means is, if you are or have been reacting to anything I've suggested, please ask yourself these three questions.

Are you reacting to my data? Are you reacting to my questions? Are you reacting to my data or questions as if they are meant to be answers?

Please know, if you've answered yes to any of these questions, it means I've so far failed to get you to personally experience the point of constellated science—questions, not answers. If so, I apologize. In truth, I cannot make this point enough. Why? Because it goes against one of the strongest tendencies in human nature—to seek answers.

This is probably why we're taught to focus on answers for most of our lives. All through school, our teachers repeatedly test us on knowing answers. It's not surprising then that scientists—and all of us really—favor a scientific method which focuses, start to finish, on finding answers.

The thing is, all you need do to see through this erroneous focus is to watch how babies learn. They focus on questions. Answers are never more than brief pauses. Ironically, at no other time in life do we learn more—or learn more quickly. But by roughly age six, we've been taught to stop asking questions and focus on answers. Sadly, with few exceptions, we all do this.

Sleep / Wake
(external opposites)

Sunlight / Darkness <> Light Intensity

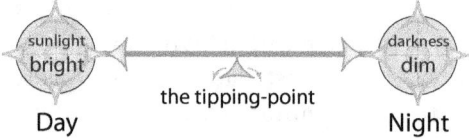

Daylight / Transitional Light <> Color

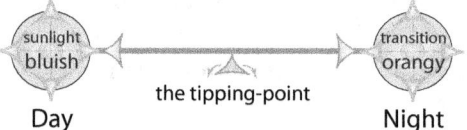

Bedroom Light / TV <> On / Off

O₂ / CO₂ Levels <> Environmental Air

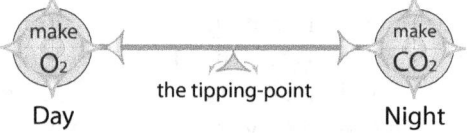

Bedroom Noise / TV <> Noise / Quiet

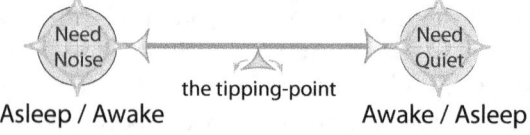

Sleep / Wake
(internal opposites)

States of Consciousness <> Awake / Asleep

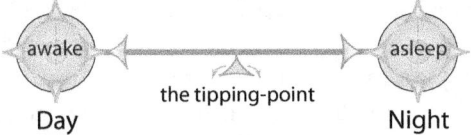

Serotonin / Melatonin Production

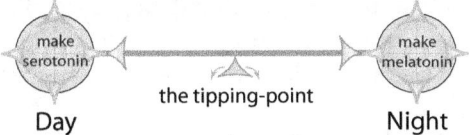

Serotonin / Melatonin <> Consciousness

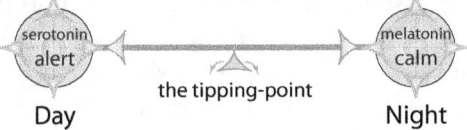

O₂ / CO₂ Levels <> Blood

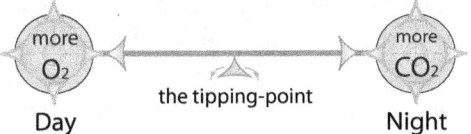

PH Acidic / Alkaline <> Blood

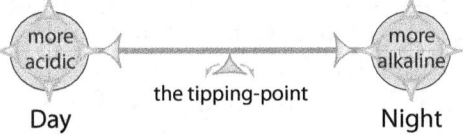

So What Have We Learned About Sleep ... ?

Finally, here are a few of the questions which emerged in me during this round of constellated science's first process—the Cartesian Process. As I've said, I'll divide these questions into the five categories I started the chapter with. The thing to watch for here will be your reactions to these questions. Curious minds are open minds. Certain minds are closed minds.

What I'm saying is, the entire point of the Cartesian Process is to open your mind. Closed-minded people can't make discoveries—and feeling certain closes people's minds. This means if you feel curious when you hear my questions, then you're beginning to see how constellated science works. However, if my questions elicit strongly negative reactions—or if you hear them as if I'm giving you answers—then it's likely, your mind is closed. And yes, after a lifetime of being pressured to focus on answers, it's normal for this to happen in just about all people. But what can it cost you to spend a few minutes wondering about the topic of sleep with me?

So does science know all there is to know about sleep? Or is there more?

Serotonin, Melatonin, and Neurotransmitters

From what I've read, neurotransmitters either excite or inhibit things—or both. My first question is, do all neurotransmitters have a hormonal, complementary opposite? For that matter, do all hormones trigger rest states for their complementary neurotransmitters? Is this what scientists mean by "inhibitory" vs "excitatory"—causing rest vs causing action?

If so, then do all pairs of neurotransmitters evoke states analogous to sleep and wake states? Is this why the neurotransmitter serotonin and the hormone melatonin function like a binary on / off switch for being asleep or awake? Moreover, does this make serotonin and melatonin the overall, master sleep / wake pair for our body / mind consciousness? Conventional science believes Descartes was wrong about this—that the pineal gland is unrelated. Could Descartes have been more right than has been realized?

Know the serotonin and melatonin I've been referring to here is that which exists—and is synthesized—in the pineal gland. Through it, our sleep and wake states are inextricably tied to the circadian cycle of day and night. At the same time, many studies say that as much as 90% of the serotonin in our bodies gets made—and is used—not in our brains, but in our guts (in the enteric nervous system). Similarly, studies show that the gut makes and contains 400 times more melatonin than the pineal gland. Unlike the melatonin levels in the brain, however—which decrease (day) and increase (night) over each 24 hour period—the melatonin levels in the gut remain steady across the entire 24 hour cycle.

These seemingly contradictory findings raise many questions. To begin with, why is it no sleep research notes how serotonin and melatonin complement each other? Why has this relationship been overlooked? Should it be, for that matter? Is it only the serotonin / melatonin in the brain which affects our sleep / wake states? Can the serotonin and melatonin in the ENS (the gut) affect our sleep / wake states too? Is it affected by our exposure to light, and by circadian rhythms as well?

Then there's the thing about digestion. To wit, do our digestive systems have states analogous to sleep / wake states? In other words, do parts of them alternatively turn off and on? For instance, do they rest in between digestive episodes, and is it the melatonin doing this? Moreover, is it possible we actually have two master sleep / wake states—one for the brain in our heads, and one for the brain in our guts? Is this why our gut must be off duty in order to get good sleep? Must both brains be asleep?

With regard to light, studies show that exposing yourself to early morning sunlight not only boosts your serotonin levels. It also boosts your melatonin levels. Is this because melatonin is made from serotonin—so if there's more serotonin, then there's more melatonin? Or can exposing ourselves to this light raise our serotonin levels and at times, not increase our melatonin levels? For that matter, is it only the light which reaches our eyes that affects our serotonin / melatonin levels? Or is it possible the light which touches our skin also affects these levels?

What about the idea that SSRI's don't actually increase our serotonin levels, but rather only fool the body into thinking there's more serotonin. What—if anything—does this do to our melatonin levels? And what does taking things which increase our serotonin / melatonin levels do to our digestion and indirectly, to our sleep? For instance, does taking melatonin hinder or help ENS functioning, and how does this affect sleep?

Then there's the relationship between norepinephrine and dopamine, yet another pair of complementary opposites. Norepinephrine is made from dopamine, making norepinephrine and dopamine a complementary pair. Also, functionally, norepinephrine is associated with concentration—whereas dopamine is associated with alertness. Does this relationship imply that concentration in some way parallels how our minds function during REM sleep? Does this also imply that alertness is the primary element in how our minds function while we're awake?

On the other hand, a side effect of medically administered norepinephrine is that it can make it harder to fall asleep. The question here is, what does it do that interferes with getting to sleep? We normally begin sleep cycles with nonREM sleep and in this state, the mind is offline. So does norepinephrine interfere with falling asleep because it makes the mind more active? Must we be mentally defocused in order to *fall* asleep?

What about the question as to whether our serotonin / melatonin levels can be too high. Can we have too much serotonin? Can we also have too much melatonin as well? And what about the ratio between these two things? Can this ratio be so out of balance that it inhibits or encourages sleep?

Also, if serotonin delivers "make something happen more" messages—and if the body makes melatonin by converting existing serotonin into melatonin—then does making melatonin decrease the messages the brain is delivering? Is this a part of how we fall asleep? Or does melatonin deliver "make something happen less" messages?

Finally, if we have more melatonin at night—and if melatonin does deliver "make something happen less" messages—then does this mean increasing the number of these "off-duty" messages is yet another part of what makes us fall asleep?

Sleep, Light, Sunlight, and Darkness

To what degree does the color of light we expose our eyes to determine how the serotonin / melatonin "affect" switch gets thrown? Is the Earth's diurnal cycle—from orangy AM transitional light, to daytime blue light, to orangy PM transitional light, to darkness—the main thing throwing this switch? Or are there other things in play? If so, what are they? For instance, does exposing our skin to this same light affect our sleep / wakes states?

As I've mentioned, scientists tell us that most of the serotonin and melatonin in the body gets made, and is used, in the gut. Is there any relationship between this serotonin / melatonin and the color and intensity of the light we expose ourselves to? Or are the levels of these enteric neurotransmitters independent of the color and intensity of the light we're exposed to? If so, what mechanism does regulate their levels?

What about bedroom light? Can varying it be used to improve sleep? Is this worth exploring more? Or is this relationship simply too complex to be useful? On the other hand, we're told exposure to sunlight usually lowers melatonin levels and increases serotonin levels. Consequently, during winter, it's normal for serotonin levels to get lower and melatonin levels to get higher—at least as compared to these levels during spring and summer. People with Seasonal Affective Disorder (SAD) are especially affected this way (Somer, Elizabeth, M.A., R.D. 1999). So this raises yet another interesting question. Can seasonally tailoring the light we expose ourselves to—including light we can't see like infrared and ultraviolet light—be an easy way to improve the overall quality of our lives?

Sleep and Air Quality: CO_2, Oxygen, Temperature

The Earth's diurnal (day to night) cycle does more than cycle the color and intensity of the light we're exposed to. It also triggers changes in the gases which green plants emit and absorb. During the day, plants absorb carbon dioxide and emit oxygen. At night, plants absorb oxygen and emit carbon dioxide. This makes me wonder if it's just a coincidence that a significant amount of CO_2 makes us sleepy—and a significant amount of Oxygen makes us more alert? Or is this actually evidence for how physiology is largely determined by environment or evolution? And is this yet one more way we are inextricably tied to the opposites in our world?

What happens if the CO_2 levels in our homes remain at higher than normal levels all day long? Does this decrease the effect night plant emissions have on our sleep? Must we experience a significant difference between awake environmental CO_2 levels and asleep environmental CO_2 levels in order to have good sleep? Moreover, can this swing be too high, and can this negatively impact our sleep? What about our lives overall?

Then there's the relationship between CO_2 and PH. PH is a factor in sleep and waking, and rising CO_2 levels cause PH levels to rise. Does this mean we can improve our sleep by doing things like increasing how many alkaline foods we eat? Would doing this lower our body's relative PH levels and improve our sleep? Or would doing this make things worse—or have no effect?

REM Sleep, non-REM Sleep, and the Nature of Rest

Much work has been done to observe the nature of sleep, especially the way we cycle between the two forms of sleep—REM and non-REM. The big question here is, why do we need two kinds of sleep? Even a cursory look shows these two kinds of sleep contain many complementary opposites. For instance, based on my work in and around personality, it seems, they each favor either the body or the mind. Here, by favor, I'm referring to the fight for dominance between the body and the mind which occurs during awake states—but which seems to all but disappear during sleep. In other words, during nonREM sleep, the body is free while the mind is offline. And during REM sleep, the mind is free, while the body is offline.

This makes me wonder if the function of sleep is to cause the mind and body to temporarily stop fighting for dominance? Is this temporary truce between the mind and body the true nature of rest? Is it also possible that being in a state of freedom—rather than in a state of rest—is what reenergizes us? Is this why we need both kinds of sleep to feel rested? Both our minds and bodies need to be reenergized?

What about how sleep scientists construct their hypnograms and polysomnograms. By this, I mean, sleep scientists admit, the structure they use is arbitrary. Some books follow the traditional structure and list the order W(ake), REM, 1, 2, 3, and 4—from top to bottom. Some list the order as W(ake), 1, 2, 3, and 4, with REM at the bottom. And one study I read listed the order as REM, W(ake), 1, 2, 3, 4—top to bottom.

For obvious reasons, this lack of uniformity surprised, and confused, me. I'd always taken the traditional order for granted and had never questioned this. After thinking long and hard about it though, I realized, it makes more sense to put the awake state in between REM and nonREM states. Hence the second hypnogram drawing I included a few pages back.

This raises more questions though, beginning with these. What is the true order for these levels of sleep? And where does the awake state fit into the whole consciousness schema? For that matter, where do daydreams, hypnotic states, alcoholic blackouts, and near death experiences fit into all this? After all, this is what sleep is—a state of human consciousness, and one we would all benefit from knowing more about.

Finally, scientists have measured inverse correlations between the levels of serotonin and acetylcholine during REM and nonREM states. During REM sleep, serotonin levels go up and acetylcholine levels go way down. During nonREM sleep, acetylcholine levels go up and serotonin levels go way down. So what does this inverse relationship accomplish, and how does it help or hinder sleep? And can serotonin and acetylcholine function similarly to the serotonin / melatonin pair, in that they act as a second REM / nonREM binary switch?

What Are Dreams and Why Do We Need Them?

As I've mentioned, a common side effect of SSRIs is that they inhibit or impair REM sleep. And this is understandable. In order to enter REM sleep, serotonin levels need to fall to very low levels. Indeed, a condition exists called Serotonin Syndrome. Here, too much serotonin causes people to suffer side effects—including difficulty sleeping.

If this is correct, then an inverse correlation must exist between the action of SSRIs and REM sleep, and by inference between the action of SSRIs and dreams. Does this mean melatonin promotes REM state dreams? And if so, what does this mean about the 20% of dreams that occur during nonREM sleep? For instance, does this mean serotonin is somehow linked to the nonREM dreams we call, "night terrors?" If so, could taking melatonin reduce night terrors? Could it make them worse?

What about the times people take afternoon naps and directly enter REM sleep? What if anything does this imply about dreams? For instance, can a deficit of REM sleep account for reversals in the normal order of sleep cycles? Or could this be a return to the kind of sleep cycles we once had as babies?

Then there's the question as to how dreaming affects our overall health. Must we dream in order to maintain our health, or is dreaming optional? And if we must dream in order to stay healthy, then can a lack of positive daydreams be tied to a lack of health as well? Can a lack of positive daydreams be tied to a lack of positive feelings about life in general?

Then there's the thing about sleeping pills and how they often inhibit dreaming. Is this why their long term use often makes people more tired over time? Sleep can't cycle normally because the ability to enter REM states is impaired? Also, a common side effect of sleeping pills is mental slowing or problems with attention or memory. Does this mean sleeping pills facilitate nonREM (body) sleep at the cost of REM (mind) sleep?

Yet another question that occurred to me during the writing of this chapter involves the content of dreams. In my work in and around human personality, I explored similarities between the symbolic nature / content of babyhood dreams and adulthood REM state dreams. Can REM state dreams be a return to the same dream state / state of consciousness we experienced as babies? And can this be why we can go to sleep with a question and wake up knowing more? In other words, does the REM dream state return us to the ultimate state of babyhood open mindedness?

Finally, there's the thing scientists posit about memory consolidation—that it occurs during sleep. So, if we have two kinds of sleep, do we have two kinds of memory? If so, do these two types of memory equate to mind memory and body memory? And if so, is this why some people are better at things like dance and sports—while others are better at logic?

Where Are We Going Next?

In the next chapter, we'll be using the same four steps to explore weight and how it changes. And of course, central to this search will be the nature of weight loss. What causes weight gain, and why is it so hard to lose weight anyway? And is there more we could be doing? Or is there something about weight, food, and exercise we haven't noticed—something which our nature places beyond our control?

This then is what we'll focus on in the next chapter—the prototype question applied to weight—what can we learn about weight loss? All I can say is, it's constellated science. So be prepared to be surprised.

One last thing. If you truly want to learn to make your own discoveries, then review my final list, then diagram and draw as many of these connections as you can. Here's the preliminary diagram I did, constellating some of the inverse / parallel relationships between a few of these questions.

Human Consciousness as Circadian States

	← nonREM	Awake	REM →
Mind / Body Dominance	body dominant / mind offline (mental atonia)	mind / body tipped, but conflicted (per mind / body orientation)	mind dominant, body offline (muscle atonia)
Dream / nonDream State	20% Dreaming - 80% nonDreaming	99.9% nonDreaming	80% Dreaming - 20% nonDreaming
Dream Character & Content	simple / more realistic	none	complex / irrational
Serotonin / Melatonin Levels	↑ Serotonin / Melatonin ↓	↑↑ Serotonin / Melatonin ↓↓	↓ Serotonin / Melatonin ↑
Acetylcholine Levels	↓ Acetylcholine	↑↑ Acetylcholine	Acetylcholine ↑
norEpinephrine Levels	↓ norEpinephrine (decreasing)	norEpinephrine (present)	norEpinephrine (absent)
Glycine Levels	Glycine (absent)	Glycine (present)	Glycine ↑ (increasing)
Blood CO2 / O2 Levels	↑ CO2 / O2 ↓	↓ CO2 / O2 ↑	↑ CO2 / O2 ↓
Respiration	stable & regular pattern, decrease or no change in rate	stable & regular pattern, stable rate	variable & irregular pattern, variable rate
Heart Rate / Blood	Pressure decreases	increased pressure compared to sleep	higher mean rate & pressure

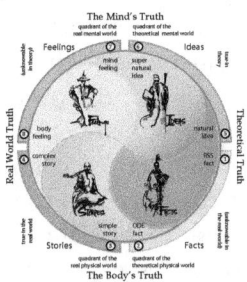

Section 2 - Chapter 11

Weight Loss
(what can we learn about . . . ?)

Why Can't We Lose Weight and Keep It Off?

Nowhere has more effort been made than in trying to discover why we keep getting fatter. Obviously, the problem can't be that we haven't tried hard enough—we make more efforts now than ever. We also know more about nutrition—and fitness—and about the risks we face. So why is it so many truly motivated people cannot lose weight?

Admittedly, science is trying to help us to understand what's making us fat. To see proof for this, simply Google, "scientific weight loss." I did and in under one second, I got something like sixty million results. At the same time, you needn't be a genius to know that most of this stuff must be nonsense. The proof? The whole world keeps on getting fatter.

So what has science been missing? Where, in fact, should they even be looking? By the end of this chapter, I promise, you'll have several new clues, including one which may be the key to it all.

At the same time, I need to tell you something right from the start. *I have no idea how to make you thin.* But before you bail, remember, what I do have is a better method. And who knows, by chapter's end, we may both be surprised by what we discover. I, for one, am curious.

We'll begin with Cartesian Process, Step One—slate clearing.

Cartesian Process ~ Step One: Slate Clearing

As I said in the last chapter, slate clearing has one purpose—to make you curious about a topic. Why? According to constellated science, you must clear your mind of the answers you think you know—answers close minds. At the same time, to make discoveries, you must feel curious. Curious minds are open minds, remember?

Slate-clearing lists are designed to accomplish both. Here, to make a list, all you need do is insert the topic into the slate-clearing question. In this case, the topic is *weight loss*. So the slate-clearing question is: "If I had access to an all-knowing *weight loss* expert, what questions would I ask?"

Now make your list.

Know it can take a bit of effort to make a list. But if you need some hints as to what might go on your list, try reading mine. One time through. Then hide my list and make your own.

Here is my slate clearing list.

- Why has the number of fat people been rising globally? (Currently, the world's population is seven billion; 1 out of every 3 is fat).
- Why is it easier for most young children to remain thin? What makes young children different? Is it mostly physical?
- Are there foods that can make it harder for us to lose weight? Or is weight gain more about how *much* we eat than how *well* we eat?
- Speaking of nutrition, can nutrition alone make us healthier AND thinner? If not both, then why not both?
- Why does eating processed foods tend to make us gain weight?
- Can eating whole foods make us lose weight?
- Will there ever be a diet that can get us, AND keep us, thin?
- Are diets actually part of what's making us fat?
- Can too much exercise make us fat?
- Is our tendency to force ourselves to exercise part of what's making us fat?
- What part does insulin play in weight loss and gain? For instance, are all obese people insulin resistant?
- Why do most people who get adult onset, type 1 diabetes tend to *rapidly lose* weight, while most people with adult onset, type 2 diabetes tend to *slowly gain* weight?
- Is there a connection between rising obesity levels and the rapid rise of type 2 diabetes? (Currently, type 2 = 90-95% of new cases).
- Why is it harder to gain new weight than to regain previously lost weight?

- Why does taking weight loss drugs often lead people to gain more weight over time?
- Does fasting affect a person's tendency to gain or lose weight? For instance, can fasting raise the chances a person will gain weight?
- Do neurotransmitter levels affect our tendency to gain or lose weight? If so, how?
- Does dieting affect neurotransmitter levels? If so, how?
- Why does appetite tend to decrease after fasting for a few days?
- If a mother diets during pregnancy, does this affect the chances her child will be obese in adulthood?
- If a pregnant mother is anorexic, does this affect the chances her child will become anorexic in adulthood?
- What part does sleep play in weight loss and gain? For instance, can getting better sleep improve our chances to lose weight?
- How do bulimia and anorexia relate to obesity? For instance, is people's obsession with—and guilt in and around—dieting / not dieting actually a mental form of bulimia?
- Are bulimia and anorexia more psychological than physical (e.g. do they come from a distorted body image & a need to be thin)?
- How does depriving children of sweets in early childhood affect their tendency to gain or lose weight in adulthood?
- What is appetite? Can we change it? Is long term change possible?
- What is metabolism? What's the best way to increase it? Can we increase it long term?
- How does the color and intensity of the light we expose ourselves to affect our weight?
- Is there actually such a thing as "happy fat" (weight gain from being content)?
- Can reducing carbs make people more able to maintain weight loss?
- Does eating less fat tend to make people fat?
- How does the awareness of time correlate to how much you eat?

Are you feeling as overwhelmed as I feel right now? I just Googled, "weight loss questions" and got nineteen million results in under .25 seconds. This raises yet another important question about the new method. At what point do you consider a slate-clearing list finished? To be honest, I'm sorry to say, I don't have an answer. At least, not a scientific answer. But what I can say is this.

I make these lists all the time. And if you recall why I make them, you'll have something like an answer. I make them to clear the slate of my mind and to make myself curious. I do this so I can make discoveries.

So is my mind open right now? Indeed, it is. And if yours is, then your list is good enough. But if not—or if you think you already know the answers to most or all of your questions—then you might want to keep writing. At some point, your list will make you curious and your mind will open.

Cartesian Process ~ Step Two: Fact Gathering

In step 2, we'll begin to explore the questions on our lists. To do this, we'll first boil down our questions to a few general categories. We'll then explore these categories, looking for any and all relevant data. And remember, in this book, we're only looking to learn the method. So I'll refer mainly to already summarized facts. At the same time, I'll omit any and all conclusions. Conclusions are answers—and answers close minds.

Know that were I doing an actual step two, I'd focus more on the facts themselves—along with the sources of these facts. This said, here are the categories I derived from my list.

- Why has more dieting and better exercise led to most of us getting fatter? What have we been missing in our understanding of dieting and exercise?
- How do age, metabolism, nutrition, appetite, and will-power affect our ability to be thin?
- What role do hormones (e.g. insulin), neurotransmitters, genetics, environmental factors, and sleep play in our struggles to lose weight?
- What part does the mind play in all of this? For instance, are conditions like anorexia and bulimia mainly responses to psychological efforts to avoid getting fat? Are they exaggerated states of mind and body? Or are they something else?

Are Dieting & Exercise Making Us Fatter?

Decades of meta-studies of diets (huge collections of facts) have all resulted in the same outcomes—over time, all diets fail at the level of the average person. And yes, there are exceptions. Some people do lose weight while dieting—a few even manage to keep this weight off (5-10 %). For the most part though, in the long run, diets do not make people thinner.

The thing is, were this the whole story, then perhaps we could say we just haven't found the right diet. But this is not the whole story. As it turns out, more dieting and better exercise has not only failed to make us thinner. It's actually made most of us fatter. Indeed, from 1980 to 2000, the percentage of Americans who were obese more than doubled. In fact, it increased from 15 to 31 percent of the population (Mann, T., 2007, UCLA). So what is going on? How is it possible that more and better efforts can result in worse outcomes—especially when these efforts are based on more and better science?

It turns out, this question may hold the key to a real weight loss solution, beginning with the interplay between age, metabolism, and weight gain.

Should We Finish Our Food or Not?

One clue to this puzzle may lie in the differences between young children and the rest of us. By this, I mean, the differences between what young children get told about food and exercise versus what we tell ourselves. We tell ourselves we must *eat less* and *exercise more*. And this makes sense. The two main factors in how much we weigh appear to be how much we eat and how much exercise we do. Ironically, normal young children—the largest group of humans to be, by nature, thin—are told the very opposite of what we are told.

We are told to eat less and to exercise more, remember? But most young children are told to finish their food (to eat more) and to stop running around (to exercise less).

Here then is one of the more intriguing observations we might make about age, weight, and the mind. And it all centers on what young children are told about eating and exercise versus what we tell ourselves. Again, young children are told they must FINISH their food and STOP running around. Whereas we tell ourselves to NOT FINISH our food, and to START running around.

What makes me think these two pairs of opposites (what we tell ourselves about eating and exercise vs what children are told) are so significant? After all, can't it simply be normal for most young children to be thin—and natural for us to get fatter as we age? In truth, to a point, being thin at a young age is normal. Obviously, it's also natural for us to gain weight as we age. But the real question is, is it natural for us to become overweight as we age? Or is excess weight gain the evidence that we're doing something wrong?

Questions. Questions. Questions. Already, we're off to a good start.

Why Are Young Children Naturally Thin?

Of course, there are exceptions. Genetics can be cruel to some of us. But for those of us whose weight falls into the bell of the bell-shaped curve, it seems, it was once natural for us to be thin.

The thing is, most of us—including most parents—worry about becoming fat. So why do so many parents worry their young children may become too thin? Is it possible these parents see thin young children as being more at risk for health problems?

If so, then why do most overweight adults see being fat as making them more susceptible to health problems? Is this mainly due to being told things like that being fat will lead to us dying prematurely? Or is our fear due to something else? For instance, does it have something to do with the fact that, in both cases (as young children and as adults), we're being told to do things which contradict our natures?

Please realize, what makes these contradictions notable is, first, that these observations are so common. They're also notable because perfect symmetries like this are uncommon in nature. This said, I'm sure—were I to ask you if you've ever worried about being too thin—in all likelihood, you'd have no memory of ever having felt like this. Over the years, I've even had a number of my clients tell me they believed they were always chubby—never thin. When I had them bring in childhood photographs, though—so we could look for evidence of this—most realized they had been no bigger than many of their classmates, most of whom they remembered as being normal sized.

This raises even more questions. For instance, were my clients' childhood worries about being chubby a premature version of what adults come to believe about weight? More times than not, my clients had at least one parent who compulsively worried about becoming overweight themselves. Did this parent's obsession with becoming fat implant this fear in my clients? Or are we so culturally afraid to get sick and die that we obsess over whatever we think may postpone the inevitable?

Ironically, a century ago, people felt the opposite feelings. Back then, people saw being too thin as putting them in danger of dying—and they saw being chubby as a protection. Admittedly, I've pretty much always thought those beliefs resulted from ignorance and poor science. But were they? Or was it true that being chubby back then made them healthier?

Is Scale Weight a Predictor of Death?

These questions remind me of one of my own more memorable realizations. This realization involved a time when I was weighing myself and within seconds, had two opposing feelings. On the one hand, as I looked down at the number on the scale, I felt afraid I'd die from being too thin. In the very next moment, I felt afraid I'd die from being too fat.

Obviously, these two things can't both be true. You can't die from being too fat *and* too thin. Nonetheless, as I looked at my weight that day, this is exactly how I felt. A moment later though, I had an aha—then laughed at my own insanity. In effect, for me, no weight had been healthy. All weights were equally deadly.

This raises yet more questions, obviously including how I could have felt this way. My first thought is that my mother died as a direct consequence of anorexia at age 48. Conversely, my brother-in-law died as a direct consequence of being grossly obese at age 59. Did these two events lead to my strange, conflicting feelings? To this day, I do not know.

What I do know is that weighing myself that day permanently decreased my fears that my weight would cause my death. In large part, this aha also motivated me to ask more weight-related questions. Over time, one of the more prominent questions has been, "do we all get programmed to connect unhealthy body weight to premature death?"

Realize that when I say my mother died "as a direct consequence of anorexia," I'm saying her body weight was a factor in—not the cause of—her death. People with anorexia have compromised immune systems. My mother got the flu, and she had no immune system to fight it off.

At the end, she was 5' 7" and 70 pounds.

She had weighed 105 pounds the day before.

Similarly, when I say my brother-in-law died "as a direct consequence of being obese," again, I'm saying his body weight was a factor in—not the cause of—his death. People who are grossly obese are notoriously difficult for anesthesiologists to correctly anaesthetize during operations.

My brother-in-law weighed over 400 pounds.

He died during a hip replacement.

Would they both have lived were it not for their weight problems? Many people would assume this to be true. I do not.

At the time, my mother had been severely depressed for several years and had been asking herself if she wanted to live. Similarly, my brother-in-law had been beaten down by years of pain in his hip, as well as by the blame he regularly and frequently heaped on himself for being fat.

So are these things taken together the cause of their deaths? I don't think it's that simple. For one thing, like philosopher, David Hume, I do not believe in simple cause and effect—life events are just too complex to be teased apart in this way. For another, were I to make a list of questions as to possible factors in their deaths, I'm sure it would be quite long. My main reason though is that the new scientific method has taught me to steer clear of mind-closing answers.

Unfortunately, the current method has yet to make this connection between closed minds and answers. Moreover, this is not due to a lack of effort on the part of scientists. They've literally been mandated to seek reasonable causes for all natural things. These logical inferences then get treated as if they ARE the literal nature of things.

For instance, my mother's death certificate states she died of "heart failure." How stupid is this? Does anyone die without having heart failure?

As I think about this last idea—science's obsession with cause and effect—many more questions arise in me. For example, it occurs to me that the current scientific method may be part of why so many people see weight as a predictor of their deaths. Can this be true? Indeed, is the current method a factor in why we generally blame our deaths on ourselves? If so, what part does the stress and worry that results from this belief play in our deaths—and in our health—and in how much we weigh?

Are Calories Real?

A second weight-related concept we often connect to predictions of death is the idea of calories. Ironically, while calories are something every weight conscious person learns to fear, few realize this idea originally had nothing to do with food, let alone whether a food would make us fat. To wit, in the nineteenth century, calories were a unit of measure in the newly evolving science of thermodynamics—the science of all things hot and cold. Strangely, at the time, scientists believed things got hot because they contained an invisible physical substance—*caloric*. This word comes from the Latin word *calor*—meaning heat. And an object's temperature was said to come from how much caloric this thing contained.

This belief then changed when scientists began to accept atomic theory. Here, heat was seen as what happened when molecules rubbed against each other and collided. More rubbing and faster collisions meant stuff was getting hotter and vice versa. And at this point, the existence of caloric became unnecessary and the idea fell out of favor.

Today, the idea that things contain a substance called *caloric* has long been forgotten But we retain the word *calorie* as a unit of measure. For instance, when it comes to weight and weight science, we use the idea of calories to measure two things—how much potential heat energy a food contains, and how much kinetic heat energy exercise creates. In both cases, the word *calorie* still refers to a measure of heat. But in the case of food, it refers to how much heat energy a food can supply if we eat it. While in the case of exercise, it refers to how much actual heat gets generated and used during various kinds of events.

Obviously, when it comes to weight, many people believe these two references to *calories*—how much we eat and how much exercise we do—are the heart and soul of weight loss. But is this all there is to it? For that matter, should we even be restricting our focus to physiological phenomena—such as how much food we put into us and how much we physically use up? Or should we be considering other factors, such as how our thinking contributes to what we weigh? Know we'll return to this question in a moment. For now, we'll stick to physiological factors.

Is Metabolism Real?

Metabolism is yet another concept we often connect to predictions of death. And in my seventh grade science class, I was taught metabolism equates to an invisible, physiological thermostat. I was also taught, we can turn this thermostat up or down based on the kind of food we eat and how much exercise we do. And that these two physiological factors—eating and exercise—then determine our tendency to gain or lose weight.

Today, we still talk about exercise as something which "burns" the calories in food. Admittedly, like the mythical physical substance, *caloric*, the image of physiological *metabolic thermostats* gives us a good way to talk about our weight. On the other hand, this can't be all there is to it. Despite all the attention we give to calories in / exercise out—we're still not thin. So what have we been missing? What can we not have tried?

More important, can it be these analogies have falsely convinced us we understand the problem? We know there is no substance in food called *caloric*. And we know we do not have literal, *metabolic thermostats* in us either. So can it be, treating these two things as if they literally exist is part of what's been preventing us from making progress?

As usual, I have no answers. Only questions. And the question which nags me the most here is simply this. Can the states I'm calling *naturally thin* and *naturally fat* actually refer to points on the real metabolic scale? In other words, can the continuum between these two poles be the real scale for metabolic rate?

I for one feel really curious about this possibility.

Why Does Diabetes Change People's Weight?

When I was young, I was fortunate to go to an extraordinary grammar school. I vaguely remember being told it was ninth academically in the country. In large part, I'm sure this was due to this school's principal, William O. Schaefer. He let boys physically fight until they had black eyes and became friends. He had famous "negro" singers inspire us with acappella—in an all white school, in the 1950's. He spoke straight from the heart, and never pulled punches—and he understood children. And he once told the whole school we were using too much toilet paper, and that if we didn't stop, we'd be doing our business in the woods.

I mention Mr. Schaefer because we all loved him. Sadly, toward the end of my years at that school, he died. He died because, in his early 40's, he got late onset, Type 1 diabetes. Somehow, they never got it under control and like my mother, he died from complications of being too thin.

Years later, I remember being scared and confused when a close friend got diabetes. Given my childhood experience with Mr. Schaefer, this is understandable. But when I asked my friend to tell me what happened, what he told me confused me. He told me he had diabetes Type 2—but with some occasional, Type 1 traits.

Me being me, I immediately launched into trying to learn all I could about diabetes. At which point, yet another pair of nature's symmetries emerged—one which may apply to what we're exploring here. I realized that while the two types of diabetes have a lot in common, nonetheless, they are also polar opposites as well.

Again, this pattern is notable for two reasons. One, because both types of diabetes are common enough to be called, fractal patterns. And two, because like having a symmetrical face, this kind of natural symmetry is rare.

Constellated scientists call the first thing—the things which these kinds of pairs have in common—their *threads of similarity*. And they call the second thing—the things they have in symmetrical opposition, their *threads of dissimilarity*. As I said, the rarity of the first thing makes it scientifically significant. What then makes the second thing important is that, in theory, what lies between these two poles defines the theoretical limits of our search. This means, with regard to diabetes, everything we can discover about it should lie somewhere between these two types.

If so, then contrasting and comparing these two natural pairs will set our search in motion. We'll start with what the two types of diabetes have in common. What are their threads of similarity?

Threads of Similarity: Type 1 & Type 2 Diabetes

- Both types center on some kind of trouble with insulin.
- Both involve abnormal levels of blood sugar.
- Both involve complications as a result of these fluctuating blood sugar levels.
- Both can be life threatening, especially over time.

- Both require ongoing monitoring and management in order to compensate for the symptoms.
- Both include the potential for unwanted changes in body weight.

Now we'll list their threads of dissimilarity—the qualities they have which make them complementary opposites. We'll begin with Type 1.

Threads of Dissimilarity: Type 1 Diabetes

With Type 1 diabetes, the pancreas doesn't make enough insulin. This lack of insulin prevents the body from moving glucose from the blood into the body's cells to be used as energy. When this occurs, glucose levels fall (low blood sugar), and the body starts burning fat and muscle for energy. This causes the person to begin to rapidly lose weight.

Untreated, this rapid weight loss can lead to serious symptoms and death. Obviously then, Type 1 diabetes has something important to tell us about weight. For one thing, our weight must be connected to pancreatic functioning and to insulin. My question here is, to what degree—if at all—can we use this knowledge to better manage our weight?

Threads of Dissimilarity: Type 2 Diabetes

With Type 2 diabetes, the body does something completely opposite. Here, the body does make enough insulin. But the cells become insensitive or resistant to this insulin. This renders the body unable to efficiently move glucose from the blood into the cells, causing our energy levels to flag. The body then tries to compensate, by making extra insulin.

Unfortunately, the higher a person's insulin levels get, the harder it is to lose weight. Insulin is anabolic, meaning, it tends to cause the body to store fat. At the same time, the more weight a person gains, the more likely he or she is to have higher-then-normal insulin levels. This then leads the person to always be hungry—and to constantly overeat.

~

Can you see why I've included diabetes in a chapter on weight loss? Type 1 diabetes causes unwanted weight loss. Type 2 causes unwanted weight gain. More important, the idea that Type 2 diabetes increases people's appetites raises some interesting questions, beginning with this. To what degree do blood-insulin levels determine appetite? For instance, people with Type 2 diabetes often have extra insulin in their blood. They also have an increased appetite. Does having too much insulin in the blood stream signal the brain to call for more food? What about Type 1 diabetes? Does having too little insulin in the blood stream cause the brain to call for less food? If so, does this then cause people to lose their appetite?

Does the Timing of Insulin Production Match Our Eating?

Another line of questioning involves the timing of insulin production. To begin with, how soon after we begin to eat does the pancreas begin to make insulin? Is there a notable lag between food going into us and the pancreas producing insulin? Is there also a notable lag between when food stops coming in and when the pancreas stops making insulin?

Also, is it common for the body to produce more insulin than we need? If so, then does this unneeded insulin signal the brain that we need more food, perhaps to prevent low blood sugar? Can this account for the times when people feel famished not long after being full? In other words, is this feeling mainly the result of having too much insulin in the blood?

Then there's the question as to why fasting causes people's appetites to shrink in days. I've done many fasts. By day three or four, I don't even want to eat. Is this simply a case of my stomach shrinking? Or can it be, no food, no insulin—and no insulin, no appetite?

What about the idea that processed foods may cause the pancreas to misjudge our need for, and then overproduce insulin? Could the evidence for this over-production be that we get tired from low blood sugar, especially after a big meal? Can this be why we respond so badly to foods like white sugar and white flour? These foods are in such an unnatural form that the pancreas doesn't know how to properly gauge them?

Finally, some researchers claim there are two other types of diabetes—the type pregnancy sometimes causes; type 1.5, and the type which can lead to Alzheimer's; Type 3. In effect, with Type 1.5, people swing back and forth between Type 1 and Type 2. And with Type 3, you must have Type 1 or 2 to get it. It's a complication of 1 or 2. If this is true, though, then why have I referred only to Type 1 and Type 2? My response? Even if there were four hundred types of diabetes, it would not matter. Using a pair of complementary opposites defines the limits of all we can know about diabetes. At any given time, a person's diabetes can either tip into the Type 1 category or the Type 2 category. Nothing else is possible.

Are you beginning to see the scientific value in using complementary pairs? Can you also see why a scientific method must employ them or risk theoretical—and real world—incompleteness?

Do Processed Foods Cause Weight Gain?

Most of us have known for years that *processed* foods have negative side effects. Nutritionally, most are inferior to whole foods. Some are outright poisonous. When it comes to weight loss though, not all processed foods are created equal. For instance, I would guess those with zero calories would tend to cause less weight gain than those with pounds of sugar. Here, "guess" is the operative word.

This said, I've wondered for years why eating these foods seems to cause so many people to gain weight. Is it that these foods are engineered to create cravings in us? It that why we can't eat just one? My intuition tells me there's no simple answer here—and that teasing this question apart would take more time than we have. So much for this mystery. However, during this chapter's data gathering process, I did notice a pattern which may shed some light on this mystery.

Oddly, I've never read about nor heard of this pattern before. And while I've not done experiments to verify any of this, I've gathered plenty of facts which support my questions.

We'll begin with a brief look at the nature of nutrient absorption.

Does the Rate of Absorption Change Nutritional Value?

Where to begin?

My story starts some years back, when I was trying to lower my cholesterol. I wanted to do this without taking a statin. So I searched for alternatives. In the process of investigating my options, I read a wonderful little book. This book was chock full of useful suggestions. One suggestion involved taking mega-doses of niacin. I'd tried this years before and had ruled it out. Mega-doses of niacin give most people terrible hot flashes, and this had happened to me. But this book's author named a company whose patented *time-release* formula made niacin flash-free for most folks.

I tried it, and it worked really well.

I mention this as, prior to that, I'd not considered the benefits of time release medications and supplements. I'd assumed the main benefit was that they were just more convenient—less to remember and all that. However, it seems the main benefit has to do with how long they stay in the blood stream. It turns out the efficacy of what you eat is directly proportionate to how long it stays in your blood. Things that stay in the blood stream longer are more potent than things which you digest quickly.

For instance, with niacin, to get the full benefit, the recommended dose is 3,000 mg. of normal release niacin. But with time release niacin, this recommendation gets cut in half—1,500 mg. gives you the same effect. Not only that, but because the dose gets cut in half, there are fewer side effects. So half the drug gives the same benefit with fewer side effects.

At the time, I thought to myself, why would anyone not take the time release version? This led me to wonder why the normal dose doesn't just wait in line to get absorbed? One answer people offer involves the idea that cells have a maximum absorption rate. Moreover, when cells reach this rate, what they can't absorb just gets discarded. When I read this, I began to look for examples, and was surprised to find one in the same little book.

In this example, the author touted the benefits of phytosterols. Phytosterols are a close chemical relative to cholesterol. But while cholesterol comes from animals and can harm us, phytosterols come from plants and can't. Moreover, it turns out that most cholesterol is absorbed in the small intestine, and that the small intestine mistakes phytosterols for cholesterol. Temporarily, anyway. So if you take phytosterols before you eat, they block the absorption of some of the cholesterol in what you're eating.

What's important to see here is why phytosterols block the absorption of cholesterol. The body absorbs cholesterol only when it passes by the special cells meant to absorb it. If something is already being absorbed by these cells, the cholesterol just keeps going. So if you take phytosterols before you eat—and if the body is trying absorb this when the cholesterol is passing by—then the cholesterol just passes by and gets excreted.

Are "Pre-Digested" Foods a Cause of Weight Gain?

Admittedly, this explanation is beyond oversimplified. This said, it points to why time release supplements and medicines work the way they do. All cells have a maximum absorption rate for whatever substances they absorb. Once reached, like backed-up toll booth lines, it takes time before more can be absorbed.

Now consider how this absorption limitation affects how the body processes food. We eat food in large part because it contains nutrients which the cells in the body absorb. For this to happen, the body must break foods down so it can get at these nutrients. Indeed, a big part of digestion involves this breaking down process. And in general, the more complex a food is, the more time this takes—and the slower it goes into the blood stream.

Realize, the idea applies to everything we take in—both physically and mentally. To learn (to absorb) ideas, you must take them in at a rate you can absorb. Clearly, this absorption limitation applies to both the mind and the body. And to see how the break down speed affects absorption, try picturing the places where the body absorbs physical food as holes in a wall—and picture the food as balls we're trying to throw into these holes.

Now imagine you're standing three feet from a hole, and the ball is small—while this hole is big. How hard would it be to toss this ball into this hole? But now say you're standing on a conveyor belt, which is slowly moving to the right. How hard would it be now? Finally, imagine the conveyor belt speeds up to 30 miles an hour. How hard would it be to get the ball in the hole now? Could you even do it?

Now consider how this explains why whole foods are generally better for us than processed foods. Whole foods pass through the body more slowly—because it takes the body longer to break them down. On the other hand, most processed foods come "pre-digested," in that they're already broken down to an easily absorbed form. Think whole grains versus white flour and corn on the cob versus high-fructose corn syrup (HFCS).

This raises an important question. Are whole foods nature's version of time release nutrition? Does their slower rate of absorption explain at least part of why they're better for us? Does it also explain why processed foods tend to make us gain weight? Are processed foods similar to normal vitamins, in that they give half the benefit with more negative side effects?

Then there are the questions I mentioned before about how the rate of food absorption affects the pancreas. Does the pancreas have a maximum rate at which it can produce insulin? If so, then do processed foods exceed this maximum rate? Or does the pancreas just compensate by speeding up? If so, then can this cause errors in the amount of insulin produced?

Finally, realize, when I say "pre-digested foods," I'm not referring to all processed foods. I'm referring only to those foods which have had their internal structure physically or chemically broken down. Typically, these foods have had their natural fiber processed or removed. Is this why studies show that pre-digested foods take half the energy and time to digest?

What About Appetite, Hunger, and Fullness?

Next we're going to explore yet another of weight science's ideas. Here, even a casual look reveals that this topic is as complex as sleep and weight. Indeed, the topic of appetite is so complex, I could easily write an entire chapter about it. But being we're not looking for answers, we'll limit our search to finding questions about how appetite affects weight.

How will we address this complexity? Obviously, we'll need to collect data. We'll also do what Einstein did—we'll use thought experiments. Later, in step four, we'll create a few snippets of logical geometry to map out the patterns we find. This time though, we'll need to add a few twists to the geometry we create. Specifically, we'll need to visually represent how appetite patterns *change over time*.

Know we made a drawing like this once before, in the previous chapter. We did it when we drew the alternate hypnogram drawing. There we charted how a pair of complementary sleep states (nonREM and REM) change over time. But in that drawing, we were able to adapt an already existing drawing, by altering the placement of the nonREM and REM states.

When it comes to appetite, though, there are no already existing drawings. So we'll need to start from scratch. Fortunately, this will give me an opportunity to show you how to use logical geometry to make "change over time" diagrams—we'll do this toward the end of this chapter. For now, we'll focus on a tool I have yet to introduce you to—the idea that *all symptoms (signs of suffering) are a compulsive replaying of what was once a normal, healthy, natural response to life*. And what could be more symptomatic of problems with appetite and weight than the two weight related conditions—anorexia and bulimia.

Anorexia, Bulimia, & Being Obsessed With Weight

As I said earlier in this chapter, my mother died at age 48, as a direct consequence of anorexia. At the time, this word didn't exist—her death certificate lists heart failure. Strangely, today, most professionals see anorexia as coming from a desire to be thin. In other words, they claim anorexics choose not to eat so as not to get fat.

They couldn't be more wrong.

I have vivid memories of my mother telling me, at age seven, to chew each mouthful of food 32 times. She'd read this made you healthier—even then, I was skeptical. But looking back, I can understand this—her advice simply fit her nature. And by this time, she'd been "over-chewing" her food for years, as she tried to get it down.

Of course, the relevant question here is this. If my mother was fighting urges to overeat—because she was worried about getting fat—then why in the world would she have trouble eating? Duh. It was easy for her to *not eat*. She only had trouble *eating*. So much for the psychological nonsense that anorexia comes from a fear of getting fat.

On the other end of the spectrum lies the condition known as bulimia. These folks have a terrible fear of getting fat—because they have trouble *not eating*. Forget about chewing food 32 times. They consume huge amounts of food. Then once they're stuffed, they work as hard as my mother worked to get food in. But in their cases, they work hard to get this food out.

The thread of similarity between these two conditions is that these folks all suffer from a gross inability to maintain a *naturally* healthy weight. Healthy weight maintenance does not require extreme measures. This explains why bulimic people suffer even when their body weight is chart perfect. And why anorexic people like my mom can be genuinely concerned about being too thin—yet be unable to eat. Moreover, if you create a continuum with bulimia at one end and anorexia at the other, we see the truth. Bulimia is a variation of obesity wherein extreme measures are employed to avoid being obese. And anorexia is a condition wherein extreme measures must be taken to not die from being too thin.

What conclusions can we draw about appetite from these descriptions? Obviously, anorexia and bulimia both cause pain in and around food and weight. They're also both long term conditions, frequently requiring physical and mental interventions. But if you recall what I've said about young children and adults, you begin to see a connection—naturally thin young children get told to finish their food and stop running around. Naturally fat adults tell themselves to not finish their food and to start running around.

This begs the question—can these two conditions be extreme versions of those two natural states? Do anorexics tell themselves they must finish their food and to not run around at all? And do bulimics tell themselves to never finish their food and to run around to a fault? Here again, we see the scientific value in creating pairs of complementary opposites. In this case, the pair creates the baseline from which we can explore all other aspects of appetite. On the one end, people feel like a little food is too much. On the other, people feel like too much food is too little.

Finally, please set aside any thoughts as to how people get these conditions. In a book on scientific method, cause should not be the focus. The focus should be on what we can learn about the nature of these conditions. Including that I've met a few people who have been both anorexic and bulimic, but never at the same time. In fact, I vividly recall a conversation I had with a thirty something year old woman who fit this description. In that conversation, she told me her personal experiences of both conditions. The part that stays with me is how differently she felt with each of these two things. When she had bulimia—she felt obese, hated herself, and felt compelled to get rid of food. But when she had anorexia—she knew she was thin, loved herself, and felt compelled to eat.

Sadly, when professionals diagnose anorexia, they focus only on logical explanations. By doing this, they miss the very nature of this condition. So can anorexics be afraid to both lose and gain weight? Yes. And to see why, think back to what I wrote in the section on scale weight. There I told a story about a day that I looked down at the number on the scale and felt both afraid I'd die from being overweight—and afraid I'd die from being too thin. Both fears had embedded in my brain.

People who have had both anorexia and bulimia often have both fears embedded in their brains as well. Afterwards, these fears can coexist during either condition. So when the diagnosis for anorexia states that anorexics can't eat because they are afraid of gaining weight, this is why. These folks have had both. But when people have had only anorexia, they know they're thin. So they try to force themselves to gain weight.

My mother, who died of anorexia in her 40's, only felt the latter. My closest friend's sister, who died of it in her 50's, also only felt the latter. Indeed, most of the anorexics I've met and tried to help in my life only felt the latter. Sadly, medical theorists rarely make knowing people's true natures more important than logic.

Here then is yet another example of how constellated science's pairs of complementary opposites clearly define things. Appetite gets defined on the one end by people who feel too little food is too much food—and on the other by people who feel that too much food is too little. Can you see how this makes it logically impossible to mistake one condition for the other—let alone for a person to have both conditions at the same time?

The current method claims you can have both at the same time.

The current method is wrong.

What About Neurotransmitters and Weight Loss?

Now we'll take a brief look at the role of neurotransmitters. We'll begin with the idea that the hypothalamus—a small organ in the center of the brain—is said to be the main regulatory *organ* for human appetite. This is not to say this organ causes appetite. It excretes hormones. And besides hunger, these hormones are said to govern aspects of body temperature, thirst, sleep, circadian rhythm, moods, sex drive, and the release of other hormones in the body.

Digging further, we're told three hormones control appetite—insulin, ghrelin and leptin. We're also told that the neurotransmitter *serotonin* affects appetite as well. Indeed, the neurons that regulate appetite appear to be mainly serotonergic—although neuropeptide Y (NPY) and Agouti-related peptide (AGRP) also play a vital role.

Then again, serotonin is also said to affect mood and social behavior, digestion, sleep, memory and sexual desire and function.

Do you see something odd here? The *hormones* the hypothalamus excretes and the *neurotransmitter* serotonin do a lot of the same things. This raises many questions, including this one. Are these hormones and this neurotransmitter more the fine print in the appetite mystery—rather than the source of appetite? In other words, when we look at hormones and neurotransmitters as the cause of weight problems, does doing this only serve to confuse matters more? Indeed, is engaging in this sort of reductionist science any different than blaming whole food groups for our weight problems? In truth, you can't change what you can't see. Thus it seems, breaking people into parts—while seductively logical—never leads to genuine understandings of the nature of our problems.

What About Will Power and Losing Weight?

Philosophers have been arguing for millennia as to whether we actually have free will. Most folks claim we do and that we make our own choices. Most philosophers disagree. They claim we each have a nature and that this nature biases our every act. And if you think about it, this pair of opposites—nature and will power—go a long way to prove they're right.

Where does free will come in then, for instance, with regard to losing weight? To see, combine the idea that we have a nature with the idea that this nature can tip. For instance, in this chapter on weight, I've repeatedly referred to the possibility that our natures may tip between being naturally thin and naturally fat. If so, then this may explain where will power comes in. Will power is what makes our natures tip.

Speaking of natures, did you know that therapists have a word for the things which do and don't fit our nature. *Syntonic* things do. Thus, they're easy to do. *Dystonic* things don't. So they're hard. This said, obviously, exploring the philosophic aspects of will power is far beyond the scope of what we're looking to do here. Why raise it then? Because will power seems to play such a big part in losing weight and in keeping it off.

How much do we need? How do we spark it into being? How do we keep it going? What do we do when it runs out? Know that this time, I actually feel bad that I don't have answers.

Again, I only have questions.

Cartesian Process ~ Step Three: Experimenting

At this point, we've reached step three—experimenting. Here, we'll look to discover facts which make us even more curious. As we do, please remember how constellated science defines facts. All facts reference three things—a specific time, a specific place, and a specific observable event.

At the same time, to create an experiment, you must be able to describe this experiment with ideas. Here, ideas are summaries of experiments, and experiments are meaningful sequences of facts. The point is, to be a legitimate experiment, it must result in facts. Thus an idea may be well grounded in science. But even a well grounded idea is not a fact.

Here is my first experiment. I'll begin with ideas—then offer facts.

How Much Does Your Mind Affect Your Weight?

I have often wondered how the mind affects people's weight? In November 2000, I explored this in an informal, online study. Here, I enlisted the help of a group of about 100 people. Then I asked them to interact with me by email for 30 successive days. I then sent them daily emails asking them to focus on a particular weight related topic for that day, and then to write back and tell me what they'd experienced.

My question was simple. Could improving a person's mental awareness in and around eating affect their weight? At the time, I had no idea there were two natural mind / body weight states. Even so, the results so intrigued me that I did a second survey in January 2001. And for those who may be interested, here's an excerpt from the flier.

> *Do you have e-mail? Would you like to explore your eating in a truly loving and gentle way? Hello. My name is Steven, and I teach people to love the parts of their lives they struggle to be present for. And this flier is an invitation to join me and the many others who, during the month of January 2001, will be trying to do exactly this with their food and eating.*
>
> *How does being more present in and around food affect people's weight? I was, and still am, curious about the full implications. In fact, this past November, I led a month long experiment in which people personally explored this. This experiment consisted of a series of simple daily assignments, each designed to restore some of their ability to remain conscious in and around eating. People then shared back with me their thoughts, feelings, and daily experiences.*
>
> *The result? A significant number of people, including me, lost weight without dieting. In my case, I lost sixteen pounds. What so interests me here is not the weight loss though. It's that my weight loss was simply the result of trying to be more present.*

Okay. I admit. At the time, I had no real clue how this weight loss happened. Saying I was more present and conscious is like saying I did something better because I did something better. This said, fifteen years later, these two surveys still amaze me. People reported so many unexpected things—including weight loss without dieting—but also more fitness with no weight loss. Is this possible? All I can say is, I'm reporting facts, not conjecture. Self-reported facts, yes. But they're facts, nonetheless.

Can Eating a 4 Ounce Muffin Cause You to Gain 3 Pounds?

Has someone ever told you they gained pounds from having eaten something small? I've frequently had people claim they gained pounds after eating a 4 oz. blueberry muffin—or a piece of layer cake. Unless I'm missing something big about the laws of physics though, this can't be possible. And yes, calories count. But this much weight? Nutritionists tell me it takes 3,500 calories to gain a pound.

The thing is, I know these folks are reporting actual changes in scale readings. So if food is not making their weight vary this much, then what is? After doing the mind / body experiments in November 2000 and January 2001, I decided to explore this question in a more formal way. A few months later, I'd created what came to be called, the Yellow Book. In it, for 91 days, people recorded their weight, eating, and level of activity once a day. They also wrote brief responses to daily morning and evening questions, all centered on various aspects of eating.

Here, the first day was seen as a baseline day. Like the other days, people weighed themselves, then roughly noted how much food they ate—and how active they were. But this amount of food and activity became the midpoints (3) of two scales ranging from 1 to 5. Thus on the food scale, 4 meant they ate a little more than the zero day, 5 is a lot more, 2 is a little less, and 1 is a lot less. Similarly, on the activity scale.

As for the format of the book itself, it was organized into three parts—[1] record keeping; [2] exploring attitudes, and [3] consciously observing present day behaviors in and around food, weight, and eating. Moreover, for those who are interested (and for those who like to see me squirm), here's a brief excerpt from the introduction. Clearly, I had a lot to learn back then. And still do.

This is the Yellow Book, a 91 day journal for conscious eating. It's called the Yellow Book because we bound the original paper versions in a yellow leatherette folder.

What exactly is this book? For one thing, it's a guided, three month journal. In it, you'll visually explore your thoughts and feelings about managing your eating and weight.

More than just a feelings journal though, you'll need to visualize what you write. Thus the Yellow Book is also a form of guided meditation, a sort of spiritual journey you can take in order to identify some of what has blocked your ability to eat consciously.

Is the Yellow Book a diet then? Actually, no. As long as you do your best to consciously witness yourself eating, you can eat whatever you want—whenever you want. And yes, some people who have done the Yellow Book have lost weight. But as best as we can tell, they lost this weight mainly because they became more conscious in and around food and eating.

Please know, the thing to focus on here is the word, "conscious." You see, unlike what many people might assume I mean—that being conscious means making better choices as to what and how much you eat—the "consciousness" I am referring to here is the "consciousness" you arrived with at the moment of your birth; your "baby consciousness."

Why "baby consciousness?" Well, think about it. Babies do not need to diet. Their bodies function fine, even when they eat imperfectly.

How is this possible? They live in an elevated state of consciousness. So by nature—meaning, without effort—they eat consciously. Translation. Because they're more conscious to begin with, their bodies process food differently than adults: more effectively and with less effort.

So how did we lose this native ability; this "baby consciousness?" Most of us lost it somewhere between age three and four. Certainly by age four, most people have become less conscious in and around food and eating. In effect, they eat without knowing they are eating.

Obviously, much has changed in the fifteen years since then. For one thing, what I was calling "baby consciousness," I now call, being "naturally thin." I mention this experiment though because of what I did learn from doing it. For one thing, I learned that daily changes on a scale DO NOT reflect actual weight changes and so, do not mean much. Here's why.

Weight Changes in Ranges, Not Pounds

When I was young, the first video game I learned to play was **Pong**. Here, you played electronic table tennis against a computer. The thing that made it fun was that the ball had genuine ballistics. By this, I mean the ball bounced the way real table tennis balls bounce—with the same unpredictable bends and spins and speed changes and such.

It turns out body weight changes in similar ways, including that these changes have their own ballistics. At any given time, weight usually varies within a range. But within this range, weight varies rather unpredictably—sort of like how ping pong balls bounce off paddles and walls.

What is a weight range? To begin with, ranges are not averaged numbers. Ranges are literal repetitions of high and low scale numbers. These high and low scale numbers represent the ceiling and floor for your current weight range. And what's important to know is that it's easier to keep your weight within this range than to push it above or below it.

Now to see what I mean by this, take a look at this drawing I've placed on the following page. This drawing documents the various weight ranges I experienced during a six year period. From it, I learned four things.

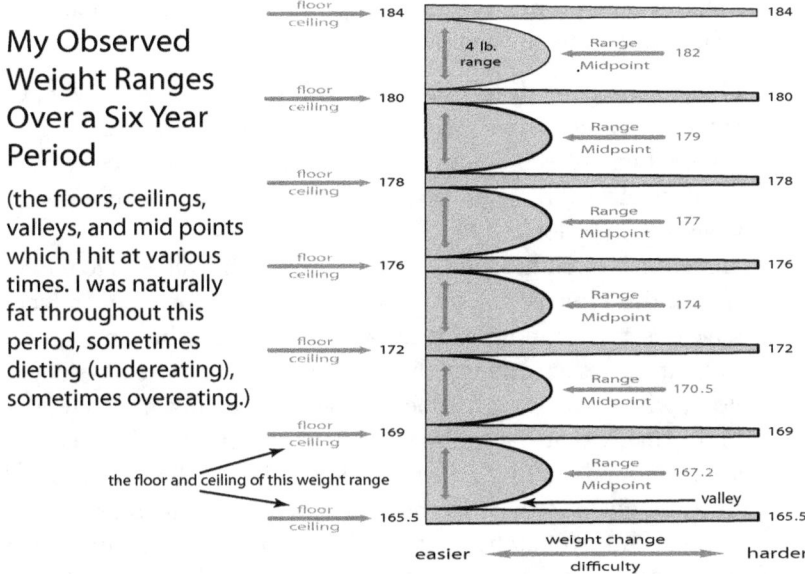

- The hardest scale weight to *move to* is to the current weight range's floor and ceiling weights.
- The easiest scale weight to *move to* is to the weights just inside the current weight range's valleys (the points just inside the current weight range's floor and ceiling).
- The hardest scale weight to *maintain* is the midpoint between the current range's floor and ceiling.
- The easiest scale weight to *maintain* is the weight just inside the current weight range's limits (the floor and ceiling).

So what are weight ranges? They're a lot like high and low tides. By this, I mean, tides normally vary between a high and a low. Moreover, high and low tides normally vary within roughly the same two points. Unless a hurricane comes—after which this range can be permanently altered.

The same thing happens to us. At any give time, our weight varies rather unpredictably—*within a range.* And this happens *even when we over or undereat.* Somehow, the body compensates and our weight remains in the current range. Know a decade of three-month weight numbers from hundreds of yellow books corroborates this—again and again. However, if we get seriously ill—or if we binge and overeat enough, this hurricane of over or undereating can push us into a higher or lower range. Then, once there—even if we then go back to our normal ways of eating—we tend to remain within this new range.

Realize, this data points to several important discoveries about how body weight varies. Here are four of these discoveries.

- One, the body tends to compensate for daily changes in what we eat and how active we are—keeping us within a range.
- Two, when we see significant daily changes on a scale, these changes are more often body weight ballistics and not lasting weight gain or loss.
- Three, once a weight range gets set, your weight will tend to remain within this range, even if you over or undereat for a day. In fact, I have observed hundreds of recorded instances wherein a person overeats and the scale weight number *goes down* the next day.
- Four, when your weight does go below the current range, it is highly likely that the following day, your weight will increase by several pounds, *even if you eat next to nothing*. I call this, "rebound." And my guess is that it's the body's way of protecting us from famine. Unfortunately, people on a diet often quit this diet when they see this kind of unpredictable weight gain. They think the diet is failing. In truth, rebound proves it's working. But not having been told this, they see the higher number and give up.

Does Sleep Affect Our Weight?

Something that has long puzzled me is why eating later in the day is said to be worse for us than eating earlier. Indeed, it seems this advice has been around forever—an old proverb suggests, "Eat a hearty breakfast, eat a light lunch—feed dinner to your enemies." Moreover, studies support this advice, and experts constantly parrot it. But why? What makes this advice true? Is it even true?

For the past ten years or so, I've paid special attention to how the time of day at which I eat affects my weight. Several interesting patterns have emerged. For one thing, while not eating for three hours before bedtime doesn't guarantee I'll lose weight, when my weight goes down—more than eight out of ten times, I've eaten nothing for at least three hours before bed. Conversely, while eating close to bedtime doesn't guarantee I'll weigh more the next day—more than eight out of ten times, when I do gain weight, I've eaten within an hour of going to bed.

This data has raised a load of questions, including why the timing of when you eat would matter. I could understand if eating close to bedtime impaired the quality of my sleep. But that's not always the case. During the writing of the chapter on sleep though, I began to wonder about

Tmin—the idea that our lowest daily body temperature occurs about two to three AM. This made me ask if Tmin is the point at which weight changes begin. In effect, can Tmin be the tipping point event wherein weight tips up or down? If so, then can eating too close to bedtime—and digesting this food while you're sleeping—raise your Tmin? Moreover, can losing weight be affected by our body temperature during the night?

Finally, all my data points to the idea that my weight varies more between days than within days. So is this pattern true for most people? In other words, does the direction of people's weight loss or gain get reset every twenty four hours? And can body weight have a circadian rhythm?

Cartesian Process ~ Step Four: Pattern Seeking

Now we'll begin the final step in constellated science's *first* process—we'll look for patterns in our data. Here our long term goal will be to use these patterns to construct new maps. At this point though, we're only searching for patterns in our data. Thus maps don't normally get created here.

I'll also introduce you to the form of logical geometry I mentioned earlier—"change over time" drawings. What makes these drawings special is they add a time dimension (Z axis) to the usual X (changing vs. unchanging) and Y (visible vs. invisible) axes. This reveals even more patterns which usually lie hidden in the data—this time, patterns of change.

Re-listing My Starting Categories

To begin with, we'll need to list the categories we've been working from. We'll then list the pairs of opposites we've discovered. To do this, I'll use a word or phrase to describe each of these pairs. Then, when we're done, I'll try to translate these words into snippets of logical geometry.

Here are our refined categories.

- The Nature of Dieting & Exercise: Why Are We Fatter?
- Age, Metabolism, Nutrition, Appetite, & Will-Power
- Hormones, Neurotransmitters, Genetics, Environment, & Sleep
- Anorexia, Bulimia & Mind / Body Connections

Cartesian Process ~ Our Working List of Opposites

Next I'll list the pairs of opposites we've discovered so far—together with a few pairs we did not have time to explore. Then we'll look for parallels in these pairs of opposites. Finally, we'll create a list of questions which we can use to further explore the nature of weight loss.

Remember though, we are NOT looking for answers here. We're only looking for patterns which lead to new lines of questioning.

Here are the pairs of opposites we've discussed or that I am about to propose.

1. The Two Core Weight Factors (how much you eat) vs (how much you exercise)
2. The Two Eating Mechanisms (Absorption vs Blocked Absorption)
3. The Two Food States (whole ~ natural) vs (processed ~ unnatural)
4. The Two Pathological Insulin States (Diabetes Type 1 ~ pathological weight loss / the body makes too little insulin) vs (Diabetes Type 2 ~ pathological weight gain / the cells can't use the insulin it makes)
5. The Two Physiological Digestive States (filling and emptying)
6. The Two Kinds of Appetite Awareness (middle road vs outer road)
7. The Two Kinds of Will Power (syntonic vs dystonic)
8. The Two Ordinary Metabolic States (being naturally thin vs being naturally fat)
9. The Two Controlled Metabolic States (dieting vs overeating)
10. The Two Pathological Metabolic States (anorexia vs bulimia)
11. The Two Weight Range Barriers (lower limit vs upper limit)
12. The Two Pathological Calorie Reactions (too few calories / can't keep losing weight vs too many calories / can't stop gaining weight)
13. The Two Weight Change Momentums (before sleep Tmin vs after sleep Tmin)
14. The Two Weight Environments (Internal ~ will power, appetite, metabolism, appetite awareness vs External ~ food, activity, light, temperature)

Designing "Change Over Time" Drawings

Now we'll use these pairs of opposites to create a few change-over-time diagrams. Hopefully, these snippets of logical geometry will make our data more accessible. Our goal will be to see if we can discover additional connections. I, for one, am curious.

To do all this, we'll first need to create a blank, food experience chart. We'll use this chart like sleep therapists use blank hypnogram charts to plot people's sleep. Of course, we'll be using ours to plot how food affects people's minds and bodies. Know we'll use four basic elements in our charts.

- Element [1] will show *how food fills and empties physical spaces in the body*.

- Element [2] will show *how this filling and emptying changes over time.*
- Element [3] will show *how the mind senses this filling and emptying.*
- Element [4] will show *what the mind does and does not sense.*

Let's start with out first element [1] how the body physically cycles between empty and full. We'll call the completely full end—"stuffed." And we'll call the completely empty end—"famished." We'll then place this continuum—from stuffed to famished—on the vertical axis. We'll place stuffed at the top, and famished at the bottom.

Now turn the page and look at the first section of this drawing (See # 1). This section explains the first element. Here, I've labeled the upper band—"my body is full (stuffed)." And I've labeled the lower band—"my body is empty (famished)." I've then divided the middle two half-bands into upper and lower bands as well. In the upper half-band, the body is filling, and in the lower, the body is emptying.

Realize this element functions similarly to a road. Here, the two middle half-bands function like a two lane highway with a dividing line. The outer two bands then function like this road's "shoulders"—one representing being physically stuffed (full), and the other, physically famished (empty).

Know that this drawing is far more complicated than it first appears. For one thing, it's fractal—and like all fractals, it contains stochastically nested elements. By this, I mean it contains a pattern which repeats at a different scales—a smaller third geometry (filling to emptying) nested within a related, but larger, third geometry (full to empty).

Now consider how the four double-headed "measures" which span this vertical continuum represent these two third geometries. Together they describe how food physically travels through the body. Here the inner third geometry—the middle road—is where balance exists. Sort of like being on the middle road in Buddhism. And the outer third geometry—the two outer roads—is where excess exists. Here we're out of balance.

Are you feeling overwhelmed? Are you wanting me to just get on with it? If you feel this, I don't blame you. We all want answers to things. Like learning the techniques fine artists use though, the more you learn, the more you can see—and do.

Next we'll need to create a way to show the second element—how filling and emptying changes the body over time. To do this, we'll treat element one as if it is a road. We'll then use a sine wave (an "S" shaped curve on its side) to plot a path down this road. And to see what this looks like, look at section # 2.

Weight Science's 4 "Change Over Time" Variables

[1] How Full or Empty Are the Spaces in My Body?

my body is full	outer road (stuffed)
my body is filling	middle road (filling)
my body is emptying	middle road (emptying)
my body is empty	outer road (famished)

[2] How Are These Spaces Filling and Emptying Over Time?

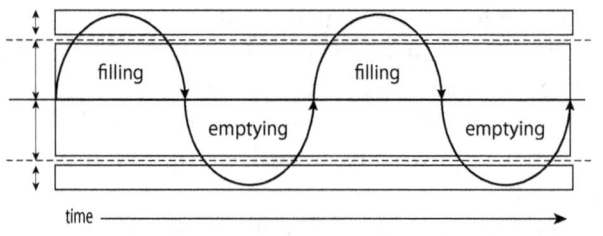

[3] Can I Sense That My Body Is Filling and Emptying?

(solid)	← I am aware of this →	outer road (stuffed)
(foggy)	← I am unaware of this →	middle road (filling)
(foggy)	← I am unaware of this →	middle road (emptying)
(solid)	← I am aware of this →	outer road (famished)

[4] Can I Sense When Filling Becomes Emptying (and Vice Versa)?

a transition dot ◉ (a change from aware to unaware and vice versa)

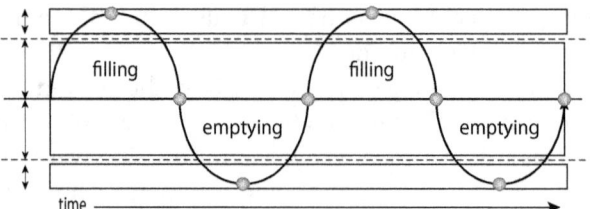

Can you see how the sine wave shaped line moves above—then below—the midline? Above the midline, the wavy line represents the body's changing amount of fullness. Below it, the body's changing amount of emptiness. Can you also see how this wavy line weaves out onto the shoulders of this road? This means, in this case, we ate until we were stuffed—then we didn't eat again until we were famished.

Truly a case of overdoing it with food!

Now we need to create the third element—a way to represent our sense of this filling and emptying. Do we realize we're getting hungry? Do we sense we're getting full? Here, solid backgrounds will represent the things we *can* sense. And foggy backgrounds will represent things we *cannot* sense. And if you look at section # 3, you'll see what I mean. Here, this road's shoulders are solid—while the road itself is foggy.

Now can you see how these backgrounds create yet another nested pair of third geometries? This "food awareness" continuum stretches from not knowing you're full to not knowing you're empty. In the case of the person in this drawing, she can sense when she's famished (very empty) and when she's stuffed (very full). But she senses little to nothing about how it feels to have food pass through her.

This particular food awareness pattern seems to be the norm for most adults. Most adults sense little to none of what occurs on the middle road. At the same time, most adults do sense hunger and fullness when it reaches extremes. And since you can't change what you can't see coming, if this is your pattern, then you'll be susceptible to over or under eating.

Finally, we need an element to represent our ability to sense transition points. Can we sense when filling begins? Can we sense when emptying has finished? Here I'm referring to sensing transition points in the two halves of the cycle—sensing changes in direction in the filling half, and sensing changes in direction in the emptying half. (See # 4).

Now consider the famous "twenty minute delay"—the idea that after we stop eating, it takes us twenty minutes to sense how full we are. Can you see how not knowing we're filling and emptying is why we experience this delay? The transition "dots" in this drawing represent the points at which we become aware— or unaware. Aware of what? That food is physically passing through us.

Okay. We now have our blank eating chart. But before we use it, a few reminders. By now, you know the big one—we're looking to discover questions, not answers. Also, keep in mind that our main focus here is to learn to use logical geometry. Finally, realize that in these drawings, we'll focus on using the third geometry to explore natural opposites.

The Power of These Drawings

Before we begin, take a moment to imagine what you might explore if you knew how to make these drawings. For example, I've used them to chart everything from changing sleep-state patterns (hypnograms) and changing appetite states to the patterns which underlie addiction.

What makes these drawings so powerful? It's simple. *They visually organize facts as they change.* This means you can chart anything you can represent with the third geometry and potentially open minds.

Want to see this at work? Just ask yourself this. Which would you prefer to be looking at—columns of statistical data or these drawings? In truth, the more you look at statistics, the more blank your mind becomes. Whereas the more you look at these drawings, the more you feel drawn in.

This explains why we're using these drawings in this step of the Cartesian Process. We're visually organizing what we've gathered in the prior three steps. This in part is what makes constellated science such an improvement over the current method. And yes, the present method does use shapes, textures, colors, and lines to represent data. However, this data is statistical. Whereas constellated science's data is by nature, visual data. And when you base shapes, textures, colors, and lines on visual data, it's like reading children's books. It makes you curious—and your drawings come alive.

The Nature of Dieting & Exercise: Why Are We Fatter?

Now let's see what patterns we can find in the data we've gathered. To do this, we'll focus mainly on our first question—can the current state of our nature influence the way dieting and exercise affect our weight? Moreover, can the state most children live in during their early childhoods—wherein it's natural for them to *lose* weight unless they're told to eat—be one end of the continuum? And can the other be the state most of us spend our adulthood in—where it's natural for us to *gain* weight, unless we fight it?

Obviously, there are times in adulthood when it does become natural for us to lose weight. Serious heart breaks. Extreme illnesses. Times of excessive stress. But again, this simply raises the same question—what makes people effortlessly lose weight in these times? Certainly it's not as simple as that they have lost their appetite. Duh! Yes. But why?

Then too, we've all known people who seem to stay thin magically. My closest friend is one of these people—he does nothing at all to keep from gaining weight. And no, he is not excessively active, nor anorexic, nor sick. Nonetheless, somehow, he effortlessly remains fit and thin. And if being naturally thin is what defines metabolism, then this makes sense.

Appetite & the Two Natural Weight Loss States

(1a) Eating in the Naturally Fat State
(most adults—and all genetically overweight children—exist in this state)

Hunger & Appetite: People in the naturally fat state will often feel urges to overeat. This means they must exert will power to control these urges or risk gaining weight.

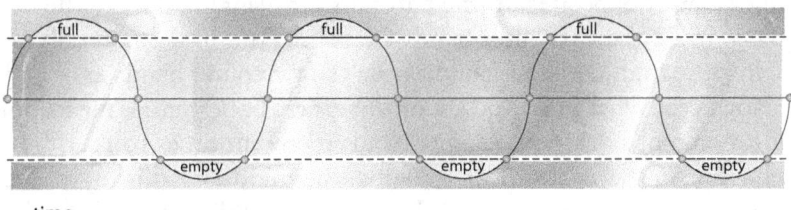

time

Most adults—and all genetically overweight children—register hunger and fullness signals only when they feel discomfort. Thus they're unaware of filling and emptying signals most of the time. This is why they often cannot control when or how much they eat. They cannot control what they cannot sense.

(1b) Eating in the Naturally Thin State
(most young children—and all naturally thin adults—exist in this state)

Hunger & Appetite: People in the naturally thin state normally don't feel urges to over or undereat. They sense changes in their bodies, allowing them to make healthy eating choices.

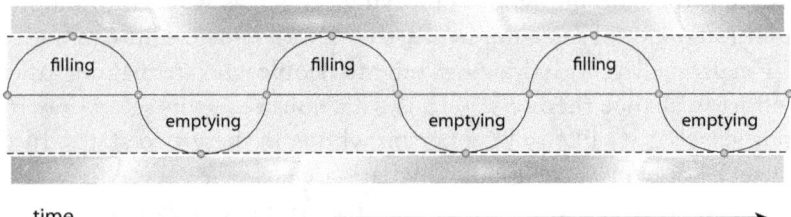

time

Most young children—and all naturally thin adults—report a mild, ongoing sense of hunger and fullness. They feel their bodies filling and emptying as this occurs. For them, hunger and fullness are gentle, ongoing changes. These ongoing changes then allow them to manage when, and how much, they eat.

Now notice the titles of the drawings on the previous page. The first—*Eating in the Naturally Fat State* (figure 1a)—charts appetite and awareness in people who, left to their own devices, tend to gain weight. The second—*Eating in the Naturally Thin State* (figure 1b)—then charts appetite and awareness in people who, left to their own devices, tend to lose weight. And when I say "tend to," I mean these drawings represent what happens if these folks do nothing to contradict their natures. No diets. No special exercise plans. No throwing up, or laxatives. No forced eating.

What do these diagrams represent? To see, pay close attention to the backgrounds of the outer and middle bands—and how they are yet another pair of opposites. In the naturally fat diagram, the middle band is foggy, while the two outer bands are solid. And in the naturally thin drawing, the middle road is solid, while the two outer bands are foggy.

Now recall what I told you about what these backgrounds mean. Solid bands represent awareness—foggy bands represent times you can't sense much. Thus in the *naturally fat* drawing, the middle road is foggy—and when the sine wave is in this band, this person senses little to nothing about what's happening to their appetite. At the same time, the backgrounds of the two outer bands in this drawing are solid. So when the sine wave passes through these bands, this person does sense what's going on.

Now take a moment to consider what this must be like. You start to eat but sense little to nothing, then suddenly realize you're stuffed. You soon sense nothing again, but then you suddenly realize you're famished. At which point you rapidly eat, then stop, when you again feel stuffed.

Compare this to what's happening to the person in the *naturally thin* drawing. In this drawing, the background of the middle road is solid, while the backgrounds of the two outer bands are foggy. This means, when the sine wave goes through the middle road, this person has an ongoing sense that they are filling and emptying. Whereas when the sine wave goes through one of the outer bands, this person registers little to nothing.

Finally realize, both drawings refer to people who are making little to no effort to change themselves. Thus what you see in this set of drawings describes what it's like to be a person who is in these two states. In the next set of drawings, we'll look at what happens when these folks try to act in ways which go against their natures. And in one case, we'll look at what it's like to be a naturally fat person who is trying to lose weight by dieting. And in the other, we'll look at what it's like for a naturally thin person to deliberately choose to overeat.

Before we do though, we first need to take another look at the sine waves and dots in these drawings.

Tracing the Appetite Transition Points

Okay. Go back to the drawings and trace the sine-wave shaped event lines. Notice how these lines cross back and forth through the three bands. Remember, we're only aware of what's happening when this line occupies a band that is solid. So when the line crosses into a foggy band, we become unaware of our hunger and appetite. And when it crosses back into a solid band, we become aware of what's happening.

Can you see where I've placed transition dots in this drawing? Remember, these dots mark the points at which our appetite awareness changed. Either we became able to sense changes in our hunger and appetite—or we lost this ability. And the dots mark these points in time.

Now focus once more on the upper drawing—the one titled, *Normal Eating in the Naturally Fat State*. Now try to recall your last meal, and in particular, the things you recall being aware of. Most adults can't remember much other than the points at which they felt discomfort. The rest remains buried in unawareness. And by discomfort, I mean the times when you felt the unpleasantness of either a too empty, or a too full, stomach.

Realize, these two experiences—feeling too empty and too full—only exist in the outer bands. What we experience in the middle band—emptying and filling—never feels this extreme. Indeed, most of what is experienced in the middle band feels rather pleasant.

Sadly, most normal adults, rarely if ever, become aware of these pleasant feelings. Young children, on the other hand, feel these pleasant feelings most of the time. This is why most young children rarely overeat—they don't like feeling full or empty. Which is why they promptly speak up whenever they begin to feel hungry or full. And why they can so easily manage—in a healthy way—when and how much they eat.

Now ask yourself how often you only notice you're hungry after you've become uncomfortably empty. Similarly, how often do you stop eating only after you've become uncomfortably full? And yes, with common sense and will power, you can make better decisions. But these decisions don't come from doing what comes naturally. Your mind is overriding your nature.

What's it Like When Our Appetite Awareness Changes?

What's it like when your ability to sense hunger or fullness changes? To be honest, explaining this may take a bit of time. For one thing, it isn't as simple as a switch getting thrown. And to see what I mean, start with this.

Whether you know it or not, your body is processing food every moment of every day. And being that you aren't eating all day long, this means your digestive system is either filling or emptying. Now think of your last meal while you try to picture what I've just told you. Picture the food passing through you—into your mouth, your esophagus, through your stomach, small intestine, large intestine, and out. Well maybe not out.

The point is, the awareness I'm talking about here is of food moving through you. At first, empty spaces fill up. Then these spaces become empty again. Moreover, at first, this happens sort of like a snake sensing a lump of food. Then things slow down and the more your body needs time to absorb things, the slower this movement becomes.

Can You Have Chart Perfect Weight & Be Naturally Fat?

At this point, I realize, I may have been unclear as to what I mean by the terms *naturally fat* and *naturally thin*. For one thing, I am not referring to whether a person IS fat or thin. A thin person can be naturally fat—all bulimics are. Here saying bulimics are naturally fat refers only to their natural tendency to do things which lead to weight gain. Given they don't prevent it. Conversely, overweight people can, at times, become naturally thin. This often happens when naturally fat people get ill or suffer a loss. And yes, in these times, their appetites decrease and this does affect their weight. But this decreased appetite is just a part of being naturally thin.

Overall the thing to realize here is simply this. These terms refer only to people's current, natural tendencies to lose or gain weight. Of course, when people do things which oppose these inner urges, their weight will be affected. And dieting and bulimia are two examples.

Does Dieting Make Us *More* Naturally Fat?

Exploring the implications of this question has been one of this chapter's main goals. Here the data clearly points to that dieting negatively impacts the current state of our nature. The current state of our nature then influences how both diets and exercise affect our tendency to gain weight. And if true, then this explains why dieting worsens most people's struggles with weight. As the Buddhists say, "that which you resist, persists." The more you try to force yourself to be thin, the more it's in your nature to get fat. Here, "dieting," by definition, is forcing yourself to eat less than you want.

By this definition, anorexics can't diet. At least, not to lose weight. Conversely, bulimics—by this definition—are guaranteeing they will only make things worse. Why? *Because they are forcing themselves to act as if they are dieting.* And as I said, resisting parts of your nature only makes them stronger. The more you try to keep food out, the more food you want.

Controlled Eating & the Two Natural Weight Loss States

(2a) Dieting in the Naturally Fat State
(does forced under-eating cause people to become even more naturally fat?)

Hunger & Appetite I want to eat more, but I try not to. I'm often hungry, even when I eat. Basically, my logical mind tells me when to start and stop eating, and I force myself to comply.

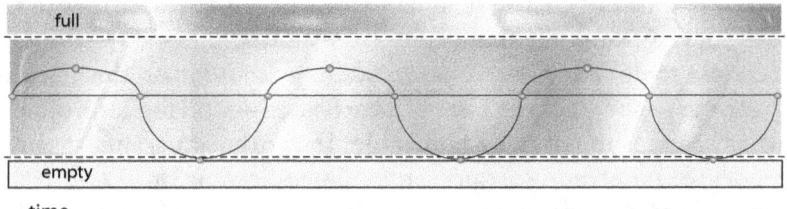

When people diet, they do many things which go against their nature. This includes forcing themselves to eat very little, even when they're hungry. Can this unnatural force be what's behind boomerang weight gain? More important, can this force permanently change a person's nature?

(2a) Overeating in the Naturally Thin State
(this is what it's like for naturally thin people to occasionally overeat)

Hunger & Appetite Most of the time, *if I pay attention*, I can notice I am filling or emptying. Occasionally, I do choose to overeat, knowing I'll end up feeling uncomfortable later.

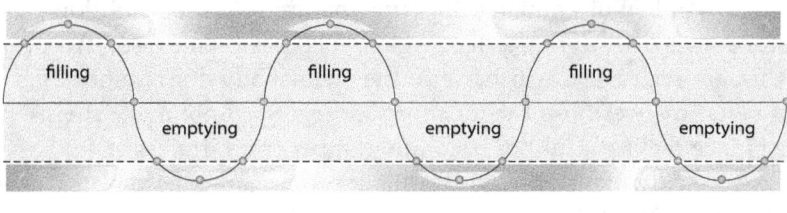

Naturally thin people have an ongoing sense of how their bodies empty and fill. For them, becoming hungry and full are *gradually* increasing sensations. Know it's these gradual, ongoing change signals which allow them to occasionally *choose* to over or undereat without permanently gaining weight.

What about ordinary dieting though? Is it really that bad? As I said, the facts speak for themselves. If you want to gain weight, diet. If you want to gain a lot, diet a lot. The question then becomes, if not dieting, than what? My answer? This book is all about questions, remember?

What About When Naturally Thin People Overeat?

What about when naturally thin people overeat? Is it okay for them to do this? Obviously, in theory, anyone can overeat—even an anorexic. Moreover, doing this creates discomfort in all people. But when naturally thin people overeat, it doesn't change their nature. So it's fine.

Why doesn't this change their nature? Mainly, it's because they remain aware. So when naturally thin people overeat, they know they're doing it and choose to do it anyway. Why would they choose to do this? Because they can't get enough of the taste of a particularly good food. As opposed to what happens to naturally fat people. They overeat because they're to some degree numb and so, can't sense fullness coming on.

To naturally thin people, overeating feels a lot like what it feels like when normal people occasionally choose to drink too much. This doesn't turn them into alcoholics. So unlike naturally fat people (and alcoholics)—who feel bad when they over indulge, to naturally thin people, an occasional excess—when it comes to food—is one of life's true pleasures.

Are Bulimia and Anorexia, Mind / Body Disorders?

I previously spoke briefly about anorexia and bulimia. Again, a good way to see the nature of these conditions is to use the symptoms tool. In both conditions, we see what experts call "signs and symptoms"—evidence of suffering. And this motivates many compassionate professionals to want to help.

Sadly, in lieu of knowing the actual nature of these conditions, most professionals turn to the ad hoc logic I call, "personal cause." According to this, anorexics don't eat because they personally don't want to get fat. This blatantly overlooks the idea that anorexics cannot make themselves eat. Hence feeling afraid of becoming fat becomes irrelevant.

Admittedly, many otherwise brilliant people get seduced by this clever-sounding logic. Unfortunately, the more science seeks these "personal causes," the less it looks for natural causes (observable evidence). In some professions, observable evidence now takes a back seat to clever, but groundless, ideas. So while folks often fault psychology as a "soft science"—because it's mostly made-up logic—it seems, they're often fine with applying the very same groundless logic to the nature of physical things.

Uncontrolled Eating & the Two Natural Weight Loss States

(3a) Bulimia as the "Unnaturally" Fat State
(this state is a reaction to forced undereating in the naturally fat state)

Hunger & Appetite: Most of the time, I feel hungry (empty). So I constantly want to eat. But when I do, I feel stuffed, then guilty and ashamed. So I force myself to purge as much as I can.

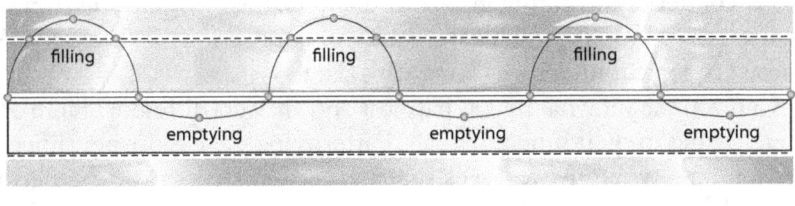

Folks with bulimia live in an "unnaturally fat state." They feel a constant sense of hunger (emptying)—all day long. This emptying feeling tells them to eat. But as they do, they cannot sense filling. Then when they suddenly feel full, they panic. Then they purge. This causes the dreaded fullness signals to go away.

(3b) Anorexia as the "Unnaturally" Thin State
(this state is a reaction to forced eating in the naturally thin state)

Hunger & Appetite: Most of the time, I am unaware whether I am filling or emptying. But when I eat, I get no joy from doing it. I rarely want to eat. But I know I must. So I force myself to eat.

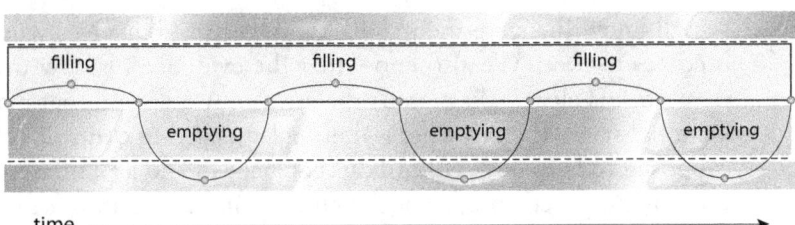

Folks with anorexia live in an "unnaturally thin state." They rarely feel empty—thus they rarely feel hungry—and this leads to them feeling false full signals. In effect, whenever they eat, they already feel full before they begin. To eat, they then must force food into what feels like a full stomach.

As for the actual nature of bulimia and anorexia, it seems to boil down to two questions. Is anorexia a state of being wherein a naturally thin person becomes so naturally thin, they become unnaturally thin? Conversely, is bulimia a state in which a naturally fat person becomes so naturally fat, they feel compelled to make unnatural (extreme) efforts to not become fat?

Haven't I Just Broken My Rule & Offered Answers?

I know. I've repeatedly urged you to focus on questions—not answers. But I've just offered you several drawings which to some might seem like diagrams of answers. My response? The drawings I've offered you here diagram patterns and comments—not answers. Here, by patterns, I mean visual representations which reveal natural connections in the data. As opposed to logical inferences about this data.

Can you see yet what makes patterns and answers different? Here one difference lies in their inherent goals. Patterns look to expose new lines of questioning—while answers seek sure and certain ways to avoid suffering. A second difference then lies in how patterns make analogies to the visual nature of things—what things look like; what they are—while answers look to logically explain why these patterns look like what they look like.

Obviously then, my drawings are pattern based. They use visual analogies to describe one or more weight related states of mind and body.

So What Have We Learned About Weight Loss... ?

If we've successfully constellated our weight loss and gain data—new lines of questioning should arise. Here, these new lines of questioning measure the value of our science. Indeed, constellated science's step four, prototype weight question might be better phrased here as, "So what new lines of questioning have arisen?" Have we succeeded? Let's look.

- We all know diets are mental efforts to control ourselves when it comes to food. We also know that the evidence for how well they work is clear—diets aren't the answer. A question that may be of help here then is, what is dieting supposed to control? Is it supposed to help us to control our behavior in and around food? Or is it supposed to control the current, internal nature of our bodies—our natural, adult tendencies to overeat?
- The big question then is, why do diets tend to make us fatter? Is this mainly a case of the Buddhist maxim, "that which you resist, persists?" In other words, the more you diet, the more you become naturally fat? Or is this but one factor in a constellation

- of negative effects? For instance, does being naturally fat push us to make poor food choices, and do these poor food choices then worsen our tendencies to become naturally fat?
- Obviously, good nutrition plays an important role in our health. Here again, the evidence for this idea is clear. But while better nutrition often does directly correlate to better health, it does not directly correlate to weight loss. To be honest, every time I read about this lack of correlation, it confuses me. I, for one, would like to know why good nutrition and losing weight don't directly correlate?
- At the same time, being in the state of mind and body I'm calling *naturally thin* does directly correlate to having a natural preference for nutritious foods. During the two decades I've explored weight loss, I've seen numerous examples for this. In other words, naturally thin people tend to naturally eat food which is more nutritious. They also tend to lose weight. But why do naturally thin people tend to eat healthier? Can it be there is something about losing weight that improves our ability to taste food?
- Current studies repeatedly show that exercise alone does not directly correlate to weight loss either. On the other hand, the more you feel a natural desire to exercise, the more weight you tend to lose. Why? Are there really only two essential qualities we need to address to lose weight—food and exercise. To lose weight, must we just address both? Or is there something else?
- Naturally thin people often feel an unforced, internal desire to exercise more. This natural desire to exercise is defined by two feelings—[1] an internal urge to exercise more, which requires [2] an internal voice telling you to exercise less. This pair of opposing feelings much resembles what young children feel when their parents (an external voice) tell them to stop running around (a counter-response to a child's internal urge to exercise). It this part of why being naturally thin works? If so, then can controlled bursts of forcing ourselves to eat more and exercise less (the defining qualities of naturally thin people) work in our favor? In effect, can we use the Buddhist maxim, *that which you resist, persists*, in a positive way?
- Speaking of resisting, what about how will powered efforts to eat less and exercise more negatively correlate to *permanent* weight loss? To wit, decades of studies show that the more you force yourself to diet and exercise, the less likely you'll be to stay thin.

Again, what makes this happen? Is it just a case of mind over matter? Is "forcing ourselves" the heart of our weight problem? Or is there something else which lies dormant in people who tend to diet—something I haven't touched on in this book?

- Young children are constantly needy. Most young children are naturally thin. This leads me to ask, what role does honestly being aware of, and acknowledging, neediness play in being naturally thin? Can it be, the more honestly needy you become, the more you'll tend to lose weight? Obviously, many people overeat when they're anxious. But can this be all there is to it?
- What is it about heartbreak that makes some people rapidly lose weight, while others rapidly gain weight? Does this somehow correlate to low-level tendencies in these people (genes, family culture, food choices) towards getting either diabetes type 1 (lose weight) or 2 (gain weight)? Can this be stress-induced diabetes?
- Some health writers posit a connection between being overweight and compulsively buying too much food. And by "too much," they mean folks who frequently throw out spoiled, uneaten food. Some even suggest, the more you do this, the fatter you get. But is this true? Conversely, is there a correlation between being naturally thin and buying just the right amount of food?
- Losing weight and maintaining this weight loss are two different states of mind and body. Can it be that both these states (losing and maintaining) have a natural version (requires minor mental efforts to do) and a forced version (requires major mental efforts to do)? If so, then how do we measure when we cross this line? Is there an easily identified sign normal folks can use?
- Something that's bothered me for years is the warnings about yo-yo dieting. I yo-yo dieted a lot in my thirties and forties—so these warnings scare me. This raises another question. Can healthy body weight management be possible only when you employ the natural versions of both of these states (losing weight / maintaining loss)? Or can it be that to remain healthy, you must fluctuate between being naturally fat and naturally thin in order to keep your nature fresh? In other words, must we continuously be challenged by weight problems in order to keep our mental and physical reactions to food and exercise alive and kicking? In other words, is there a healthy version of cycling between naturally thin and naturally fat? Or does repeatedly cycling between losing and gaining weight destroy the body's natural ability to try to

maintain body weight homeostasis? For instance, does this push us towards getting diabetes type 2?
- Must people whose genetics make them vulnerable to obesity be obese? Or is there something they can do to offset their natural urges? Clearly, obesity is an extreme version of being naturally fat. So can inducing a form of anorexia—like surgical interventions—be helpful? Or do these interventions merely create a larger equal and opposite reaction—a greater internal tendency to be obese?
- Likewise, is there a genetic component which pushes people towards bulimia? Obviously, with regard to food and exercise, people with bulimia are unnaturally fat. But this state is largely mental—being unnaturally fat is an attitude. So is bulimia mostly a mental condition while anorexia is mostly physical?
- Many people believe that to keep our immune systems active and healthy, we must get minor illnesses (the flu, colds, etc). If true, this would apply to all people, both those naturally fat and naturally thin. But this raises a question about health in general. Does gracefully tipping back and forth between wellness and illness (and between being over and under weight) lead to the greatest health? If so, should we deliberately make efforts to cause this to happen?
- Finally, my biggest question remains the one I briefly mentioned about metabolism. To wit, are the naturally thin and naturally fat states of being the true measures for metabolism? In other words, is saying a young child is *naturally thin* a more scientifically descriptive way to say they have a fast metabolism? And is saying most adults are *naturally fat* a more scientifically descriptive way to say they have slow metabolisms? If so, then this one idea may hold the key to our learning to manage our weight. I, for one, feel hopeful when I consider this.

Onward and Upward. Is This Beginning to Make Sense?

Okay. So here we are at the end of our second Cartesian process chapter—our chapter on weight loss. Is constellated science beginning to make sense? I certainly hope so. And yes, we've got a lot more to talk about. The new method has a lot to offer. But at the very least, I hope you're seeing the good in exploring it further.

In case you're feeling iffy though, let me remind you of something. By now, you've seen myriad examples of how questions open minds. Unfortunately—living in questions, not answers—goes against human nature. Even for scientists. Indeed, as I wrote this chapter, I myself repeatedly felt pushed to offer you answers. And on several occasions—as I looked back over my work—I realized I'd inadvertently done this.

The thing is, as soon as I rephrased these sections as questions, they flowed much better. Clearly, scientific explorations voiced as questions feel better to my mind. In the next chapter, we'll try this again, only this time we'll explore deafness. Hopefully, at some point, you'll be able to use the new method on your own.

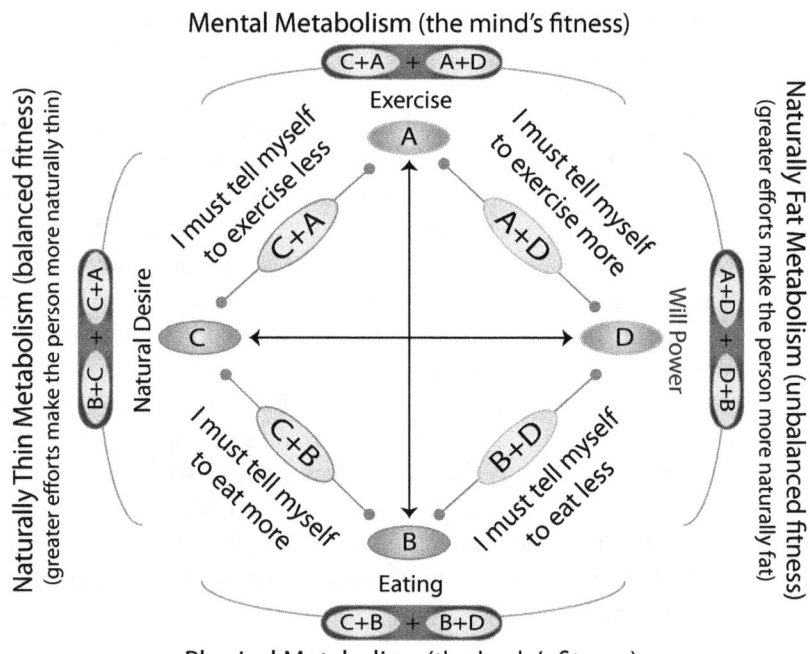

Finally, in case you're interested, here's a *preliminary* map of weight loss. Pay close attention to how the logic in this drawing unfolds in an unbroken sequence. Unlike the current method, nothing here is ad hoc. Try doing this with the current method. But please, don't blame yourself when you fail.

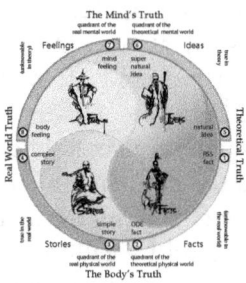

Section 2 - Chapter 12

Deafness
(what can we learn about . . . ?)

Are There Also Two Kinds of Deafness?

When I was younger, I spent two years in a relationship with a woman whose parents were deaf. And while I can still picture us laughing as she taught me to sign, "fart," what I remember most fondly were the people. Her mother, her mother's friends, the Deaf community in general—they all treated me so well. To this day, I wonder if this was due to how much Deaf people and people with Asperger's have in common. Our bluntness. Our social discomfort. Being treated by normals like we're fragile.

Years later, the six month old son of a couple I knew suddenly became deaf. And at one point, the couple asked if I'd go with them to a famous New York City deafness clinic. They wanted to explore the possibility of their son getting cochlear implants. And as I waited for them to pick me up that day, I remember struggling to make sense of how this terrible thing could have happened.

In that hour or so before they came, I urgently constelled everything I'd read. Every study. Every medical intervention. Every psychophysical observation. Every scientific interpretation. Then at some point, as I sat there, several important connections emerged.

We'll be exploring these connections in this chapter.

I mention these two stories because together, they greatly affected my life. This includes the fact that these stories are a big part of what's prompting me to write this chapter. Here, the deafness in the first story is the kind most people observe at one time or another—the kind which involves significant, but partial—and often slowly progressive hearing loss. Whereas the deafness in the second story is a kind most people will never hear of—the kind which involves a sudden transition from normal hearing to complete and often permanent hearing loss.

For reasons I'll explain later, I'll call the first kind of deafness (gradual onset, partial loss), Type 2. And I'll call the second (sudden onset, total loss), Type 1. And in this chapter, I'll be juxtaposing these two kinds of deafness to create a baseline pair of complementary opposites. My hope is that exploring this baseline pair will open the door to some new lines of questioning about deafness. And about hearing in general.

I also hope to impress on you the importance—in all scientific research—of employing baseline pairs. Certainly, working from baseline pairs is one of the major upgrades in the new method, for several reasons. One—they create a scientifically sound way to define terms. Two—they describe clear and observable limits for a research topic, insuring "scientific completeness." And three—they give us broad, yet manageable starting points from which to begin our research. This includes the present baseline pair, which we'll use to help us to create our slate clearing lists.

Cartesian Process ~ Step One: Slate Clearing

As I said in the previous two chapters, slate clearing has one main purpose. We do it to make ourselves curious about a topic. According to constellated science—to make discoveries, we must open our minds. And curious minds are open minds, remember? Moreover, since answers close minds (and in doing so, inhibit curiosity), we must do something to set aside the answers we think we already know.

Slate clearing lists convert these answers into questions.

As for where we'll begin, all we need do is insert our topic into the prototype, slate-clearing question. In this case, the topic is *deafness*. So the slate-clearing question is: "If I had access to an all-knowing *deafness* expert, what questions would I ask?"

Now make your list.

Know it usually takes people time to come up with a satisfying list. Moreover, if—while making your list—you get stuck, then try briefly reading mine. One time through. Then hide my list and make your own.

Here is my slate clearing list.

- What defines the sense of hearing and is all hearing *auditory*?
- What defines deafness (the lack of hearing)? Is all deafness the lack of auditory stimulation? Or are there other qualities which hearing people use which may be lacking in deaf people?
- For instance, what about mental hearing—the "sounds" of the things which exist only in our minds. How is mental "hearing" different from auditory hearing? And is there mental deafness?
- Do deaf people—including those born deaf—also have mental hearing? If so, how does this hearing compare to the mental hearing of hearing people?
- Do deaf people hear in ways hearing people do not? For instance, to what degree do deaf people hear through bone conduction? Do they become more sensitive to sound transmitted through bone conduction than hearing people? And do they develop other forms of auditory synesthesia (overlap between what's left of their auditory sense and their other senses)?
- Do all of our senses include some degree of synesthesia—or are there clear boundaries between our senses? For instance, to what degree does what we see create "anticipatory" sound in our minds? How much does this anticipatory hearing combine with auditory sound to create our overall experience of hearing? And how does anticipatory hearing differ between hearing people and deaf people? Also, can we be deaf to anticipatory hearing?
- Conversely, how does being deaf affect peoples' remaining four senses? For example, hearing people use sound to locate things in time and space. How do deaf people locate things in time and space? And do they by nature develop anything to compensate for their decreased auditory spatiotemporal ability?
- Do deaf peoples' experiences of the other four senses differ from that of hearing people—or is there no appreciable difference? For instance, taste is in part, auditory, and in part, visual. Does the lack of auditory sensation alter a deaf person's sense of taste, for example, when eating raw celery? Do deaf people have food preferences associated to being deaf? Do deaf people make better cooks?
- Our ears hear mainly by converting acoustical energy to electrical energy but also in part, by bone conduction. By what process does the brain convert these two forms of sound to words and speech? Moreover, to what degree does context alter what we hear? And to what degree does context change what this sound means to us?

- How does the mind intelligently isolate one sound within a group of sounds—for instance, one familiar voice within a group of voices? Is this skill entirely mental? Or is it physical as well?
- To what degree does the volume and spatiotemporal information associated to a sound allow or prohibit us from isolating a sound?
- When a deaf person's hearing gets augmented (e.g. with hearing aids), sometimes their hearing immediately improves. But sometimes it doesn't. What causes this difference?
- If a child is born hearing—but goes deaf before learning to speak, to what degree can cochlear implants allow this child to develop "normal" hearing? Should a child get both ears implanted?
- Experts claim there is a season within which to learn to hear and speak language. What causes this season? Is there a season within which to learn to hear things other than language?
- Why do most people react strangely when hearing a recording of their own voice? What accounts for this reaction? Is this mental vs auditory hearing?
- Does living with a deaf person alter the hearing of a hearing person? Does this affect the hearing person's other four senses?
- Does focusing attention on a sense like hearing cause the body to develop more physical ability to hear? Vice versa?
- To what degree does teaching deaf people to "imitate normal" (imitate how hearing people speak and interact) cause their social deficits to mirror those in people with Asperger's?
- To what degree does brain plasticity allow or limit late onset, language learning? Does the brain develop differently in a person born deaf? Does it repair itself differently?
- How much of what we hear can be attributed to pattern recognition as opposed to actual auditory functioning?
- How does being deaf affect deaf people's sleep? For instance, do deaf people dream differently? Does the silence help them to sleep? What wakes them up in the morning?
- To what degree is deafness genetic? Does deafness alter neurotransmitter balances / levels? And are deaf people more vulnerable to certain diseases than hearing people?
- What about the idea that at some point, babies become unable to recognize phonemes not included in their native language (e.g. New York babies vs Tokyo babies). Is this loss due to injury? Is it a kind of deafness? Or is it merely a lack of education?

Cartesian Process ~ Step Two: Fact Gathering

We're now ready to begin step 2—fact gathering. Here, we'll explore the questions on our lists, by collecting data. To do this, we'll boil down our slate-clearing questions to a few general categories. We'll then explore these categories, looking for any and all relevant data.

At the same time—and this is important—we'll omit any and all conclusions. Answers close minds—and we want to remain open. Also, please remember, we're mainly here to learn the method. So rather than listing details and sources, we'll populate our categories with already-summarized data.

Keep in mind though that when you do your own step two research, you'll need to include full-blown lists of individual facts and sources.

This said, here are the categories I derived from my list.

- The Physiology of Hearing: How Do We Hear?
- Sensation and Hearing: Is Sensing Sound, *Hearing*?
- The Mind and Hearing: Is Understanding Sound, *Hearing*?
- Deafness and Hearing: Do Deaf People Hear?

We'll start with the physiology of hearing—the physical design.

[1] Physiological Hearing—How Does it Work?

For the purposes of this book, I'll be calling the young boy who lost his hearing, James. And to me, one of the more amazing things to come out of my attempts to help James involves what I believe to be an important new line of questioning. This new line of questioning involves a key physiological aspect of how we hear. And like all things in our world, this line of questioning opens the door to many other lines of questioning.

Now to understand this key, we'll again turn to psychophysics. In this case, we'll look at yet another rather interesting parallel between technology and our bodies. Here, I'm referring to a surprising commonality between our inner ears and the condenser microphones used in world class recording studios.

How the heck did I make this connection?

To begin with, I owned and worked in a recording studio for a good portion of my twenties and thirties. To this day, I remain involved at the level of serious hobbyist. More importantly, I taught classes in audio engineering in my late twenties. This means I am familiar with the basics of pretty much all sound recording technology—including the peculiarities of human hearing in relation to sound reproduction.

Realize, one could easily spend a lifetime and still not be expert in these basics. This is after all an attempt to comprehend and reproduce an entire human sense. Fortunately, you will not need to learn much to grasp what I'm about to tell you. Beginning with a brief review of what experts tell us about how the ear works. Hopefully, I'll be able to do this clearly and without needing to use too much techno-babble.

To begin with, our ears have three chambers, starting with the outer ear. This first chamber is a physical horn which gathers and focuses sound. Here, the word *sound* refers to how out-in-the-world physical impacts cause vibrating pressure waves to travel through air. This is somewhat akin to how slapping the surface of the water in a pool causes sound to travel through the water.

These vibrating waves of air pressure enter the outer ear. They then cause a miniature, skin drum head to vibrate at the barrier between the first chamber and the second—between the outer ear and middle ear. This drum head then causes a series of three interconnected, miniature bones to carry the sound from one end of the middle ear to the other. And these patterns of vibrations ARE the message we hear.

Also, because these bones are designed sort of like a lever, they leverage (amplify) what we hear. This increases our ability to hear quiet sounds. Then at the back end of the middle ear, the last bone vibrates a smaller drum head. So between the leverage of these bones—and the changes in scale of the parts (big end of the outer ear horn to the little end; big drum head to little drum head), our ears intensify and focus the sound.

On the other side of the little drum head (called the "oval window") lies the third ear chamber—the inner ear. And this part of the ear is a doozy, to say the least. It's complicated, and serves a lot of purposes, and it's not completely understood. But this is the place where the magic happens.

To begin with, there's the part of the inner ear that has nothing to do with hearing—the semi-circular canals. This part consists of three fluid filled tubes. Here, as the fluid in these tubes moves, the brain uses this movement to sense where we are in X, Y, and Z space. This amounts to having a personal gyroscope in each of our ears, including that they give us our sense of balance.

Know we'll talk more about the semi-circular canals—specifically, how they interface with the brain—before we end this chapter. But for now, we'll move on to the hearing part of the inner ear—the cochlea.

Strangely, this second part of the inner ear also consists of a set of three fluid filled tubes. But in this case, the tubes function more like "conversion tubes" than balancing tubes. By this, I mean these three tubes convert the middle ear's mechanical energy waves to electrical energy waves. This electrical energy then gets carried deep into the brain by a major bundle of nerves.

Once there, the brain does what the brain does—assigning spatiotemporal and personal meanings to what's been heard. In this way, the inner ear acts as an interface between the physical world and our minds. In a moment, we'll look more deeply at how the inner ear converts mechanical energy to electrical energy. But before we do, here's a grossly over-simplified diagram of what we've just spoken about.

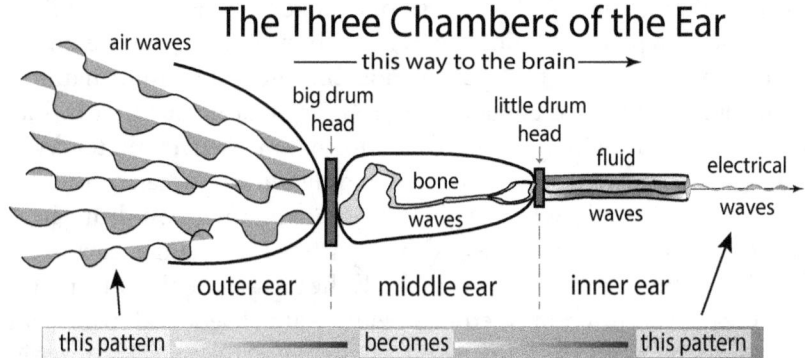

What is Hearing?

As I mentioned, at one time I taught audio engineering. I also worked in a recording studio for years. Obviously, part of what I had to know back then was how microphones work. This knowledge is what ultimately led to my asking the right questions when it came to what happened to James. Strangely, I've not read a single reference to what I'm about to tell you. We'll begin with this.

First, did you know that all sensations—including those we hear—are just us sensing patterns of change? So when we say we "sense" something, what we're really saying is that we've become aware that something is changing. Once sensed, then our brains search our data banks, looking for patterns of change we already know. If we find one, we know what we're sensing. If we don't, we try to learn the new pattern.

The point is, this process—recognizing patterns of change in what we sense—is what turns sensation into information. Thus information is actually just recognizable patterns of sensed change. It turns out this idea is the critical concept in understanding hearing and deafness. And to see why, we'll need to see how ears turn sensation into information.

Know I only realized how this works that day when I desperately wanted to help James and his family. What did I realize? That ears function like microphones and vice versa. For one thing, like ears, mikes convert mechanical energy to electrical energy. Here, microphones "hear" sound waves and convert this vibrating air to an analogous pattern of vibrating electrical energy.

This same conversion—from mechanical to electrical—is what happens in the inner ear. No coincidence, most deafness—including many things which are not seen as deafness, such as auditory processing disorder—is due to problems in this conversion process. The critical part involves realizing that it's the pattern of change that matters. And if you understand this part, you understand a lot about hearing and deafness.

Begin with what I told you a moment ago—that information is recognizable patterns of change. The thing is, it turns out that in most cases, the medium which carries this information doesn't matter. What matters is that our sensory organs—including our ears—can sense these patterns of change. So whether they're occurring in air, on drum heads, in radio waves, or in ear bones, it's the patterns of change that matter. Why? Again, because these patterns ARE the message—the information.

My point? Different mediums—air, water, electrical wires, and nerves—can carry the same patterns of change. And since information is recognizable patterns of change, different mediums can carry the same information. The ear simply copies physically changing patterns of air to analogous, changing electrical patterns in our nerves. Indeed, this same conversion process is at the heart of all of our senses. It's how we interface with our world. All five senses convert something physical to something either electrical or chemical or both.

Of course, at the end of this process is a big ball of electrochemical wiring. And so far, despite lots of theories and scientists, no one actually knows how this big ball learns and remembers. This aside, what I'm about to tell you is the most important part of all, at least when it comes to how James became deaf. Are you curious yet? I certainly was.

The Ear Needs Power?

It turns out the critical question here is one few people ever ask. The question? Do our ears need power in order for us to hear? Admittedly, when I ask people this, most get glazed-over eyes and feel confused. So if you just did, don't worry. Most people never get taught that the ear needs power to function. Then again, most people never get taught how microphones work either.

To begin with, as I've told you, microphones move information from one medium to another. Here patterns of moving air get changed into patterns of moving electricity. What's important to know here is that there are several ways microphones make this happen. Ninety five percent of all microphones used in recording studios do it in one of two ways.

- Dynamic mikes (these mikes generate their own power).
- Condenser mikes (these mikes need a power source).
- The ear is a condenser mike.

And in case you're interested, here's a diagram of a typical condenser mike.

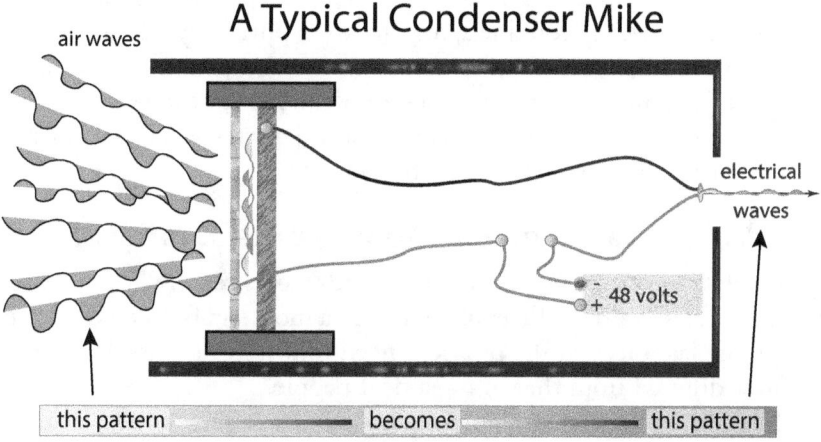

Where Does the Ear Get Its Power From?

I realize what I just told you probably means nothing to you. As I said, most people know little to nothing about how microphones work. If you do, then I apologize. But if not, here's the five cent tour.

A normal condenser microphone is literally a two-layer sandwich of electrically charged material. Here, the inner layer is rigid, while the outer layer vibrates and bends. This means, when changing patterns of moving air press against the flexible outer layer, the distance between this layer and the inner layer changes as well. And because one layer is positively charged and the other, negatively, when the distance between layers changes, an electrical quality called *capacitance* changes with it.

In this way, patterns of moving air get converted into patterns of moving electricity. The thing is, in order for this conversion to happen, these two layers must be electrically charged. As I said, condenser microphones need a power source to function. No power. No electrical output. The thing to keep in mind here then is that this holds true, *even if the microphone is physically perfect.* No power to the two layers. No sound comes out of a perfectly good microphone.

As I sat there that day, I thought back to the tests they'd done on James so far. I'd asked to see the results of these tests, as hearing tests generate the same kind of information audio engineers look at when they evaluate their equipment.

Then, as I thought back on James' test results, two things jumped out at me.

- One, they could find nothing structurally wrong with James' ears. As far as they could tell, his ears were physically perfect.
- Two, despite this lack of structural damage, the measurable levels of hearing in both of James' ears equated to the residual noise of the test equipment. Translation. James had absolutely zero measurable hearing at any frequency.

What Was Different About James' Deafness?

Okay. Can you see what was wrong with James yet? If yes, that's great. But if not, don't worry—I'll explain it all in a moment. Before I do, I need to add one last piece to the puzzle. I need to briefly describe how James' deafness differed from that of most deaf people.

First, based on testing, James had what I'm calling, Type 1 deafness. He *suddenly* suffered a *total* loss of hearing—*with no structural damage* to account for this loss. This is like plugging perfectly good headphones into a functioning receiver, but failing to turn the power on. In other words, James' equipment appeared to be fine. And everything looked like it should work. But he heard no sound. So something else had to be wrong.

This differs markedly from what most deaf people's deafness is like. Most deaf people have what I'm calling, Type 2 deafness. These folks suffer from a *slowly* increasing, *partial* hearing loss, *which does include structural damage*. Moreover, due to this damage, while they still hear, what they hear sounds nothing like what normal hearing people hear. Nor does it resemble normal sound with the volume turned down. Rather, it sounds more like what you'd hear were you to poke holes through the speakers in a pair of headphones, then put a towel over the ear cups.

This is why, when people with Type 2 deafness speak, they often sound odd—like parts of what they're saying are missing. Parts are missing, the same parts that are missing in their hearing. Moreover, since people learn to speak based on imitating what others sound like to them, the deficits in their hearing lead to deficits in their speaking.

A similar thing happens to hearing people when they're listening to loud music on headphones and someone tries to talk to them. The person with headphones on tends to respond rather loudly. Why? Because they're imitating the volume of what they've been listening to. People with Type 2 deafness do the same thing. They try to match what they say to what they hear. Hence when they speak, they tend to sound odd.

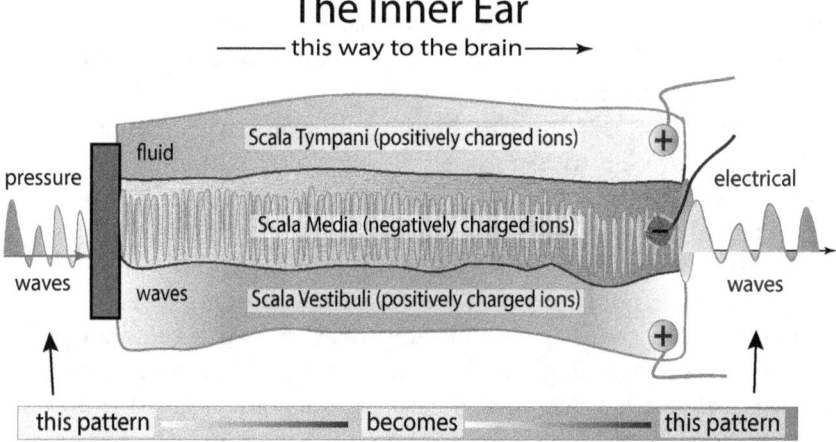

So What Caused James' Deafness?

Okay. What I'm about to suggest is not an answer. Rather, it's the contents of my current line of questioning. Moreover, I have done no direct experiments. Nor am I an expert in deafness. Nor do I have a desire to become these things. This said, I'm going to present you with some rather interesting clues. These clues strongly support my line of questioning. I'll then leave the experimenting and such to those more qualified.

Hopefully, I'll make at least a few scientists curious.

To begin with, part of what I looked for that day was for any stories involving sudden, total hearing loss, with no apparent damage. What I came across were a number of stories where drugs caused this to happen. Some drugs even cause permanent hearing loss. In fact, this group of over 200 medicines even has a name. They're called *ototoxic* medicines.

Hearing loss caused by an ototoxic medicine tends to develop quickly. It also affects—and sometimes damages—the inner ear. And while hearing usually returns after discontinuing the medication, as I said, sometimes the hearing loss becomes permanent.

The group of ototoxic medicines includes things as common as aspirin—when taken in large doses (8 to 12 pills a day). Also nonsteroidal anti-inflammatory drugs (NSAIDs), such as ibuprofen and naproxen. Also certain antibiotics, such as gentamicin, streptomycin, and neomycin—as well as certain drugs used to treat high blood pressure and cancer.

Know I am NOT suggesting that James' deafness came from taking these drugs. In truth, I have no idea whether he took any of these things or not. My point is that symptoms similar to those James had can often be traced back to medications which affect the inner ear's power source. Moreover, experts tell me this power comes from something called *mechanically gated ion channels* involving sodium (positively charged) and potassium (negatively charged) sources.

Did you get this? Ototoxic drugs can interfere with these ion gates and thus, with the electrical polarization necessary for hearing. Translation. They can turn off the inner ears' power.

Finally, for those who may doubt the inner ear actually has—or needs—this power, consider the following.

In an article published in ***Nature Biotechnology*** in November 2012, scientists announced they have developed a chip which can draw power from the ear's power supply without interfering with normal hearing. In fact, this article mentioned that the ear can generate up to 100 millivolts of electrical potential. Considering the size of the inner ear, that's a lot.

Now if you put this together—the idea that the inner ear is the functional equivalent of a condenser mike, and the idea that condenser mikes need power—a connection emerges.

To wit, the inner ear has three tubes of alternately polarized, electrically charged fluids which function similarly to the two layers in a condenser mike. More important, if you remove the power source from an otherwise perfect condenser mike, the mike will output nothing—no sound whatsoever.

My question is, could this be what happened to James? Did something cause the sodium and / or potassium ionization refresh cycles in his ear to stop? If so, did this, in effect, shut down the power to his ears? Did something turn off his ear's power switch?

Could James' Deafness Have Been Healed?

Of course, the more important question is, could knowing any of this have helped? My answer? Based on that ***Nature Biotechnology*** article, I have to wonder. To wit, if something shut off the power to James' ears, is it possible something could have turned it back on? If not, then could supplying power to his inner ears from an external source—such as a battery—have functioned like a pacemaker for his ears?

To be honest, I well up with tears when I consider that there might have been a way to help James. After all, I made this connection—between the inner ear and condenser mikes—not long after he became deaf. Moreover, because James was almost two when the decision to get cochlear implants was being made, there was a justified urgency—he had yet to speak a single word. Because of this, his parents decided to get both of his ears implanted.

My point?

In order to get cochlear implants, you must destroy the structure of the inner ear. Thus for James, there never will be a way to test this connection—James no longer has inner ears.

This only points to how urgently we need the new scientific method. And how having it might improve our world. Where's my proof?

I discovered this connection in a single hour.

Why Do So Many Parts of the Ear Come in Threes?

As a total aside, my curiosity has me sidetracked right now. So knowing me, I need to give in or risk continuing to feel distracted. My question? Why do so many parts of the ear come in threes?

To wit, there are the three chambers (outer, middle, inner), the three bones in the middle ear (the malleus, the incus, the stapes), the three fluid-filled tubes in the inner ear, (scala tympani, scala media, scala vestibuli), the three fluid-filled tubes in the semi-circular canals (the horizontal, the superior, the posterior), and the 3 drums heads in the middle ear (the ear drum, oval window, and a smaller one I haven't mentioned; the round window).

My point?

Are all these threes just a coincidence? Or are they yet more examples of sensory tritinuums (the fourth geometry)?

Is Signing the Same as Speaking?

I mentioned at the beginning of this chapter that a girlfriend taught me to sign. Not much. Just a few signs. Mostly the vulgar stuff. I mention this as the most memorable part of that event had nothing to do with learning to sign the word, "fart." What it had to do with was how I acted after I learned to sign this word.

I acted like a three year old who had just fallen in love with a word.

I repeatedly signed "fart"—over and over and over again.

This continued until my girlfriend laughingly explained to me that to her, signing a word WAS THE SAME AS SAYING A WORD. And that deaf folks feel the same way. So signing with my hand and fingers was the same thing as repeatedly speaking this word to her.

My point?

It had never occurred to me that signing is not a substitute for speaking. Signing IS speaking. I was so biased towards seeing my world—the hearing world—as the only world, that this had never occurred to me.

This hit home even more when, not long after that, I took my son to a play. And as we sat there, before it began, I showed him a few of the words I'd learned to sign. At one point, I noticed the two women seated next to me were both trying to hold back laughter. And as I turned to look, the woman next to me leaned in and whispered, "we know how to sign."

I had just signed the word for "sexual intercourse."

[2] Hearing as Sensation: Is *Hearing*, Sensing Sound?

All sensory conversions from the physical world to our minds end up at the brain. Amazingly, imagining physical experiences can cause the reverse to happen as well. For instance, when scientists have people picture themselves doing physical acts, their bodies physically respond as if they are doing these acts. This means we have sensory feedback loops in us, in that our physical-to-electrical conversions (physical sensations getting converted to electrical information) occur in the opposite direction as well (recalled electrical information gets converted to physical sensation).

Realize that despite all this talk about the ear, we have yet to define what we mean by "learning to hear." Perhaps if we combine what we've said about the ear with what we've discussed previously about the goal of a scientific method, we'll have a good place to start? Here, the goal of a scientific method is to discover meaningful patterns of connections which describe the nature of our world. Thus we could say that learning to hear is *becoming able to use sound-based, sensory conversions—to grasp and communicate meaningful patterns of connections—which describe the nature of our lives and our world.*

Now take a few minutes and try to notice what I've *not* included in this definition. Know my choice to not include this thing is every bit as important as what I've chosen to include. Can you see it yet? I've made no mention of the medium in which these communications occur. Thus despite assumptions by hearing folks that hearing is limited to sensing sound waves in air—clearly, hearing is much more than this.

For instance, what about sounds in the mind? Aren't they also part of hearing? Or should we limit hearing to what happens in air? Obviously sound travels in other things as well. What about sounds which travel through wood, metal, or water?

To be honest, I'm feeling a bit overwhelmed. Turns out defining hearing is harder than I thought. Perhaps if we set the philosophical aspects aside and try another thought experiment, we can get a better sense of what we need?

Try this. Try picturing a group of hearing people having a conversation. Now picture a Deaf woman standing next to them. Now ask yourself what she can hear. Anything at all? Now ask yourself what she can understand.

When we think of deafness, we normally picture someone who can't sense sound. Indeed, the word "auditory" shares a root with the word audible—capable of being heard. The thing is, the sounds in your head are also audible—but only to you. You can't record them with a digital recorder, nor reproduce them with any mechanical device.

So does this mean the sounds in your mind are not actually sounds? If not, then what do we call them? What you're hearing in your head? Sound-based thoughts? And what about lip reading? Is lip reading, sound-based communication? I just tried it, and I can't do it without imagining sound.

What about a Deaf man who hears sounds in his head as he reads your lips? Is he, "hearing sound?" What about the sounds you hear in your head as you read these words? Can we even read without hearing the sound of the words in our heads? I just tried it, and I can't do that either.

Even reading braille must involve some variation of this kind of sound—the kind which is inaudible to the rest of the world. It seems then that even trying to define "sound" is difficult. What does it mean to say we are hearing sound?

A further complication comes when we include the need to understand what we have heard. For example, when a friend tells you, "I hear you." Aren't they saying they "understand you" more than anything else? What about when we tell someone, "you're not hearing me." Are we telling this person he is literally deaf? I think not.

What about when we lose our temper and shout, "are you f-ing deaf?" Aren't we assuming these people hear us—so we're trying to insult them into understanding our words? My point is, it seems that *deafness*—as in, the lack of hearing—has a lot of meanings as well. So while we often equate deafness with the inability to hear physically transmitted sound, we also equate it with the inability to transmit and receive clear interpersonal communications.

Perhaps this is why we used to call deaf people, "deaf and *dumb*?" And yes, originally *dumb* meant unable to speak—being mute. But for many years now, being "deaf" has implied you're "mentally deficient." In other words, if you're deaf, you're dumb!

Fortunately, constellated science includes something which inhibits this kind of prejudice. It states that before we can scientifically explore something, we must first identify this thing's *sine qua non*—its essence. In effect, this means, before beginning a scientific discovery process, we must first define our terms. This includes big picture definitions such as the words *deafness* and *hearing*.

What do these words mean? Surprisingly, both emerge from a learning process. In other words, to define hearing and deafness, we must first define what it is we learn.

So what do we mean by "learning to hear?"

A working definition might go something like this.

Learning to hear is "becoming personally able to recognize, understand, communicate, and recall meaningful, *audible*, interpersonal *sensory* patterns." Moreover, to be comprehensive, this definition must include sounds audible only in our heads—sounds which do not rely on functioning ears. In addition, for this definition to be accurate, it must parallel what it's like to learn the other five senses. Certainly all of our senses involve learning to sense change in physical things. They also involve learning to sense change in the things we sense in our heads.

If we then add to this the interactions between hearing and the other four senses, then the main difference between each of our five senses is where they originate—in different, physical mediums.

Interestingly enough, recent observations of early childhood development point to this being true. In fact, in the August 2011 issue of Psychological Science, they describe a study that shows how all newborns are synesthetes. Translation? At birth, the five senses of newborns overlap—and continue to overlap until their brains learn to tell them apart. In other words, in babies, the five senses are not separate senses.

Can you see yet why I'm putting so much effort into defining "learning to hear?" *Deafness* and *hearing* are my baseline pair. As such, they must have threads of similarity and dissimilarity. Here the main similarity is that they both involve learning to hear. And their main dissimilarity involves learning to sense vibrations in airborne sound.

As I think about it, perhaps we should take another look at the three ear drawings. Together, they represent the physiological essence of hearing. Based on these drawings, then, what is the essence of learning to hear? Can it be it's simply becoming able to *convert, recognize, understand, communicate, and recall electrochemical analogies for sensed patterns of sound*?

If so, then obvious differences in mediums aside, does it ultimately matter which part of our bodies we use to discover these patterns? Admittedly, we lose access to much sensory beauty when we lose the use of our ears. The loss of musical experiences alone is a terrible loss—as is being able to hear our children's first words. But do these losses mean we hear nothing—that we're dumb—that we're not hearing?

If hearing is limited to sensing sound, then yes. But if not?

These questions raise yet another line of questioning. Can it be that, even in adulthood, *hearing* overlaps our other senses? A few current, cutting-edge experiments on blind people involve substituting other senses for seeing. Indeed, can the main differences between our five senses—at the pattern recognition level of learning—be solely which medium we sense? And can each of these styles of sensation lead to the same patterns of information in the end?

Finally, can the way we treat the five senses—as five physically separate things—be biasing how and what we understand about learning to hear? If so, then perhaps what we call *hearing* is actually something simpler. Perhaps it's just the act of taking someone—or something—in?

What Does Philosophy Have to Do With Hearing?

Admittedly, most scientists would complain about what I've just presented—specifically, that it's more philosophy than science and therefore doesn't belong in this book. However, according to constellated science, scientific practice and scientific philosophy are but two sides of one coin. Indeed, scientists from Aristotle's time through to Einstein called themselves, "natural philosophers"—philosophers of nature. And many scientists continued to identify this way well into the twentieth century.

The thing is, early in the twentieth century, scientific purists began to rail against doing this. They claimed philosophy can't be measured and so, has no scientific certainty. Therefore, it has no place in the everyday practice of science. The real question here is, can we scientifically explore anything without including philosophy? For instance, can we even define *hearing* without it when this word can mean so many things?

Obviously, a real scientific method must include and require clear and certain definitions. Indeed, perhaps the best arguments for this were made by the early twentieth century philosopher, Ludwig Wittgenstein. Arguably, he was the most acerbic philosopher of all time, in that he openly—and with great intelligence—mocked any and all attempts to define, and communicate with words. When someone spoke to him, he'd use the Socratic method. But he'd use it not to teach them but rather, to humiliate them into seeing how unclear their words were.

The thing is, bad attitude aside, Wittgenstein was right. Our inability to define words has crippled our attempts to know and talk about our world. This problem only intensifies further each time we try to scientifically explore things like *hearing*. Fortunately, one of constellated science's main goals is to reconcile separations like scientific knowledge from scientific philosophy, giving us legitimate ways to define "big picture" assumptions.

In truth, even when they deny it, scientists make these big picture assumptions. How could they not? For instance, if we don't make assumptions about what we're exploring, then how can we know where to begin? Moreover, nowhere do we do this more than with our assumptions about our five senses.

Strangely, I've neither heard nor read of a single, serious, scientific effort to define any of these senses, including *hearing*. So what is *hearing*? Is it mainly sensing airborne patterns of sound? And can we call this *hearing* if we don't understand what we've heard? Also, am I right in calling what we hear in our heads, *sounds*? And is noticing these "sounds," *hearing*?

If so, then are mental noises like tinnitus and other kinds of ringing in our ears also *hearing*? For that matter, can a deaf person even get tinnitus? I, myself, do not know. Or is asking this question an oxymoron—because deaf people, by definition, can't hear? Obviously, in a chapter on deafness, we must raise these questions. And obviously, in order to address these questions, we must employ scientific philosophy.

In our next section, we'll explore these questions. Specifically, we'll ask, should our explorations of hearing be limited to sensing physical sounds? Or must we assign a meaning to what we're hearing before we can call what we're doing, *hearing*?

As we look, please try to keep the following idea in mind. If you already have an answer, then your mind is closed. Rather, please do try to keep your focus away from answers, and try to focus on discovering new lines of questioning.

Is Noticing Sounds With No Meanings, *Hearing*?

Now turn your attention to the bottoms of the three ear drawings once more. Can you find the pointers at the ends of the horizontal bars? These pointers point to how and where physical patterns of sound get converted to analogous patterns of electrical energy. Clearly, this conversion process is the physical sine qua non of learning to hear.

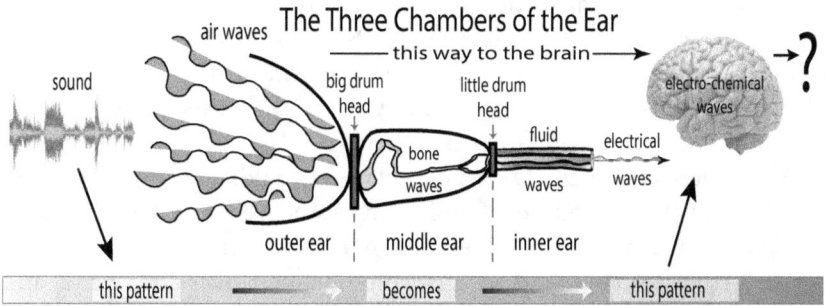

The thing is, given this process is working, is the result of this physical conversion process, *hearing*? Or is there more to *hearing*? For instance, must our minds assign meanings to these sounds in order to call what we're doing, *hearing*? Let's look.

To begin with, in the earlier three drawings, I failed to include the beginning and end of this conversion process. In the latest drawing, I've included these parts. To wit, on the left side, you'll see the source of what you're hearing—a physical world sound. And on the right side you'll see the end point of this journey—electrochemical brain waves in your brain.

In a way then, this drawing is a more complete version of what we do when we learn to hear sounds. We connect out-in-the-world sounds to our mind's ability to mimic these sounds. At the same time, we also connect these mental sounds back to what's out in the world. And still, this doesn't completely describe the process of learning to hear sounds.

The question we must now ask is, does "learning to hear" involve assigning meanings to sounds? Know that in step three, we'll explore this question in depth, when I describe my experiments with replicating what young children go through as they learn to hear, read, write, and speak a language. To do this, I chose what I saw as a difficult language—modern Greek. And like all languages, Greek isn't just sounds. It's also meanings.

Admittedly, in the years I studied Greek, I never reached even the level of a normal Greek five year old. What I found however was that learning to *hear* a language is far harder than I'd ever imagined. Certainly, Greek words have meanings. But first, you have to learn to hear the word-sounds. And at first, I couldn't hear some of these sounds. Here, I encountered another kind of deafness; what I'm calling, *language-specific deafness*. And to see what I mean by this phrase, consider the following.

How Hard Is It to Learn to Hear Greek Letter-Sounds?

Have you ever thought about how the letters of alphabets represent letter-sounds? And that when we speak, read, or think of the letters of an alphabet, we "hear" these letter-sounds in our heads? The point is, all normal children must learn to hear letter-sounds as they begin to learn a language. Thus, as I began to study Greek—I focused on learning to write and pronounce the Greek alphabet.

What happened next, I never expected—I couldn't hear some of these sounds. A few letters sounded familiar. But some were so foreign I literally couldn't hear them. I'd ask a Greek speaker to pronounce a letter, and when they did I'd just go blank. And this would happen, over and over again.

These failures led me to ask a question I hadn't previously considered. When people can't pronounce a letter-sound or word, is this actually a kind of deafness? For example, take the well-known struggles native Japanese speakers have with the English sounds "L" and "R." Forget about English speakers trying to hear Japanese.

Admittedly, part of this problem may lie in the complexity within alphabets. For instance, take English. The International Phonetic Association says it has 26 letter-symbols and 44 letter-sounds. These sounds include 20 vowel *sounds* for the five English vowel *letters* alone. Which is why most English dictionaries say the letter "A" has five "A" sounds. Then again, some say four and some say six.

Compare this to the 24 letter-symbols of the Greek alfávito, where none of the vowels have multiple sounds. Nice. On the other hand, while Greek has only five vowel *letter-sounds*, it has seven vowel *letter-symbols*. Here, two Greek letters use the "i" sound, and two use the "e" sound. Can you imagine what it's like to learn to spell Greek words? And yes, officially, there are subtle differences in how people say these letters. But in the real world? Even Greek people admit, not so much.

Remember, all I was trying to do here was to learn to hear the sounds of these 24 letters. Unfortunately, with several letters, I couldn't hear the *letter-sound* no matter how hard I tried. Add to this that while the Greek alphabet uses some of the same written *letter-symbols* English uses (for example, A, B, E, P, and X), several of these *letter-symbols* use different *letter-sounds*. For example, the Greek *letter* "P" uses the English "R" *sound*.

Worse yet were the Greek letters for which English has no equivalents. Here, it took me a month or so to learn to write them—and far longer to learn to hear them being said. For instance, take the Greek letter ψ (psi). I don't even know where to begin to describe this letter-sound. It starts with something like saying "psst" when you want to get someone's attention—only without the "t" at the end. Then you add the sound of the English word, "sigh," onto the end.

Putting it together, it ends up sounding something like, "pssssigh." Unfortunately, the troubles with this letter don't stop there. There's also the placement problem—no English words begin with this letter-sound. However, a few Greek words do (e.g. the words for bread and fish). This led me to question how placement affects our ability to hear. For instance, does placing this sound at the beginning of a word make an already difficult-to-hear *letter-sound* harder to hear? It certainly did for me.

What about when there are multiple sounds? Does the order of these sounds matter? Indeed, can reordering sounds in an unfamiliar way create a kind of deafness? Keep in mind, when it came to my struggles with hearing ψ (psi), there was nothing wrong with my ears. Moreover, this letter-sound was being spoken at normal volumes.

When it came to hearing this *letter-sound* then, whenever someone said it, my mind went blank. This raises the question, does this kind of mental blankness qualify as a kind of deafness? In reality, the sound of this letter was perfectly audible to any Greek speakers within hearing distance. I know. I asked. But when it came to me, I just could not hear this sound. I literally heard no sound in my head *until after the person finished saying this letter-sound.*

Know that over time, I did learn to hear this sound. Not only that, but I also fell in love with hearing it—and with speaking it, as well. In a future chapter—the chapter on learning—we'll discuss this rare kind of "falling in love" learning. We'll also discuss how it differs from the more common, painful, rote-learning experiences most of us endure.

At this point, we're left with two main questions. One, to what degree is "learning to hear," mastering the ability to stay conscious in the presence of a sound? And two, how exactly do we measure this mastery?

In Cartesian step three, I'll explore these questions in more detail. For now, let's move on to our third topic—understanding what we hear.

[3] The Mind & Hearing: Is Understanding Sound, *Hearing*?

This brings us to our third category—learning to understand the sounds we hear. But before we dig in, let me ask you. Can you see what makes this topic different from the previous two?

First we looked at [1] *the physiology of hearing;* how the body's physical equipment functions when it comes to hearing.

Then we looked at [2] *how we learn to sense sound itself;* how the mind learns to use this physical equipment.

In this third topic, [3] *understanding what we hear*, we'll look at how the mind assigns meanings to what the physical equipment senses. Here, we'll begin with a few more comments about my experiences learning to hear the Greek letter ψ (psi).

Recall that when I first tried to hear this Greek *letter-sound*, my mind would go blank. And that the harder I tried to hear it, the more blank I became. Moreover, since a blank mind is an empty mind, I literally became unable to hear this sound. Thus, it seems my attempts to hear this sound would momentarily render me deaf.

Realize, I am not saying I sensed nothing when the speaker spoke. I was vaguely aware I had heard some kind of noise. The thing is, hearing noise is not the same as hearing sound. Sound has a meaning. Noise does not. And to see what I mean, try doing the following thought experiment.

Imagine you are in my living room, sitting across from me. Now imagine we've said our hellos and we're about to talk. Suddenly—and without warning—I shout a word or phrase you've never heard before. How would you react?

Most people get programmed—early in life—to freeze and go blank in response to loud noises. As I think about it, this reaction may even give us an evolutionary advantage. Consider the times when a mother suddenly yells at a crawling baby about to touch something hot. If the baby freezes, he's saved from what could have been a painful burn.

Whatever the case, what turns a sound into noise is that you hear it when your mind is blank. And if a sound has a sudden onset, most times, this empties your mind.

Know there's even a simple test you can do to determine if what you heard was sound or noise. After a sound finishes, try imitating what you think you heard. If you can, it's sound. If not, it's noise.

Realize, this ability—to imitate what you hear—is actually a good way to know if you've heard it. If you can imitate a sound, you have. But if you can't, then you haven't. This test even works on things such as baby talk and song melodies. If you've heard these things, then you should be able to do a fair job of imitating them.

Why am I calling a blank reaction to a sound, a kind of *deafness?* Two reasons, really. One, a blank mind doesn't retain enough of a sound to review—let alone imitate—what occurred. Two, it also can't assign a meaning to what occurred, and meaningless sounds ARE noise. This is why, when I first tried to hear the Greek *letter-sound* ψ (psi), all I could hear was noise. And why I kept hearing only noise until I stopped going blank in the presence of this sound.

So—when I finally became able to hear this sound, I also became able to easily imitate this sound. But I also became able to connect a meaning to this sound, *as I was saying it*. Before that, the best I could do was to recall this meaning after I was done saying the sound. And this separation between sound and meaning was also proof this sound was still just noise.

Must We Assign Meanings to Sounds to Hear Them?

Some might now ask why I'm claiming I had to *learn* to simultaneously imitate the sound AND to know its meaning as I said it. Why both? After all, I intellectually understood this letter-sound's meaning long before I heard this sound. The thing is, "learning to hear a sound without going blank" and "learning to understand what it means as it's being said" *are two separate skills*. And prior to learning this second skill, I more resembled a two year old repeating a sound than an adult conveying information with words. What's the difference?

When children learn a language, they begin by learning to imitate the sounds of words. Only later do they realize these sounds also have meanings. Indeed, were you to watch a child learning to attach a meaning to a word-sound, it's pretty amazing. Each time this happens, the child gets pleasantly surprised. And this surprise permanently integrates this meaning within this sound.

To see how this works, imagine you're trying to teach an eighteen month old to say the word, "truck." You get a toy truck, hold it up, and simultaneously repeat the word "truck" slowly. The child then tries to connect these two things, and when he does, he gets excited. At what? At having made the connection between the physical toy truck and the word-sound, "truck."

Now take a moment to consider what this implies about learning to hear. Given you have properly working ears, it seems, learning to hear actually takes two steps. Step one is learning to hear a sound as sound, not as noise. Here the test is being able to reasonably imitate this sound. And step two is connecting a commonly agreed upon meaning to this sound. Here the test is to point to this person, place, thing, or event—while at the same time, making this sound.

Is "Sound Without Meaning" a Kind of Deafness?

But if hearing requires both these skills—and it seems that it may—then this raises yet another question. Does sensing a sound which lacks a meaning equate to another kind of deafness? Is this the kind of deafness a mother refers to when she asks her teen, "are you, deaf?"

For instance, say someone addresses you with an unfamiliar word or phrase. Do you call this experience, "hearing, but not understanding?" And does hearing these words without knowing what they mean turn these words into noise? I wonder. What about my claim that "sounds without meanings" are noise? And is hearing a noise, hearing? Once again, I think we need more information. So perhaps it's time for another thought experiment. Try this.

Imagine being introduced to someone with an unusual sounding name. Now imagine trying to repeat this name back to this person to see if you actually heard it. Now try to imagine doing this and at the same time, trying to picture the person. In effect, what I'm asking you to try to do is to say this person's name without a mental translation step—without the need to separately tell yourself what this name sound means.

When I'm introduced to someone who has an unusual name, I usually go blank. I then ask the person to repeat their name—then I try to say their name back to them. Realize, when I try to say this name back to the person, this only tests for the first skill—can I can imitate this sound? If I can, then I do the same thing the child in the story learning the word truck does. I push myself to connect this name sound to this person's face.

The point is, connecting this person with their name sound requires a second skill. Indeed, I find that if I don't do this, then most times, I can't recall the person's name. Why? Because I've failed to attach a meaning—them—to their name. And when this happens, it happens because the sound of this person's name is still just noise to me.

Is There More Evidence Hearing Takes 2 Skills?

Since we're still on Cartesian step two—fact gathering—we should look for more evidence for this claim. And what's coming to mind right now is an anger awareness workshop I attended some years ago. I tell this story as it offers more evidence for how sounds and their meanings affect us separately. I also tell it as an example of how hearing requires the simultaneous synthesis of both.

As this workshop began, the leader asked us to find a person we didn't know and hadn't spoken to. He then asked us to sit, facing each other, without saying a word. He then asked us to try to sense what our partner was like as a person. Then, after four or five minutes of doing this, he asked us to silently look over at our partner again.

We were then told to try to think of someone whom our partner reminded us of. We were then told to keep this to ourselves, while briefly describing to our partner what we'd sensed about who they are. Finally, our partner had to tell us how much of what we'd said was accurate.

How well did I do? To be honest, I can't recall much. Some parts of this event I remember—some parts are long gone. What I do recall though is that I'd sat across from a woman about thirty years old. Yet despite her age, she'd reminded me of my paternal grandmother.

This grandmother was a typical Italian *nonna*—extremely nurturing, gentle, and kind. And to me, for some reason, this young woman reminded me of her. Moreover, I'd attached my grandmother's personality to this woman without ever realizing I'd done it. And this had happened without us having exchanged a single word.

In part, I tell you this story to point out how much sound can change the meaning of what we hear. So did things change when we finally spoke? In large part, they did. In fact, what I heard so conflicted with what I'd assumed this woman to be like that for the rest of that day, I found myself trying to reconcile this difference. My eyes had told me one thing—the sound of her voice, quite another.

The problem, of course, was the assumptions I'd made which unknowingly included what her voice would sound like. And when the actual sound of her voice didn't match, it contradicted these assumptions. This same thing can happen when you meet a manly man who surprises you with a high pitched voice. Or a tiny woman with a gravely voice. Or a nephew just entering puberty who randomly uses both voices.

In each case, what we hear contradicts what we see.

Now consider how, in my story, my eyes had created a meaning. The woman's voice then contradicted that meaning. In effect, I could not connect what I saw and what I heard. And this points not only to how much hearing depends on both sound and meaning. It also points to how important it is for these two things to connect.

Sound, then meaning.

Two steps to learn—but one to play back correctly.

One last thing.

Think back to what I mentioned earlier about what we mean when we say someone's "hearing us." We don't usually mean they're hearing only the sounds we're making. We mean they're understanding what we're saying. Conversely, in an argument, if we insult someone by shouting, "what are you, deaf?," we're not saying they can't hear the sounds of our words. We're saying, they don't get our meaning.

Here again, we see support for my claim—that to hear, we must hear both sound and meaning simultaneously. So should we be calling it *deafness* when we miss either step—or hear these two things separately? At this point, I must confess—I feel overwhelmed by all these questions. Indeed, before writing this chapter, I never realized deafness was so complex.

At the same time, I see this emerging complexity as yet more evidence for how well constellated science does its job.

Imagine how much this method might help you to discover.

Have We Gathered Enough Evidence Yet?

Have we gathered enough evidence yet for the idea that learning to hear takes two steps? The short answer is, we'll have enough when we can use what we have to clearly define learning to hear. Moreover, if this sentence sounds like yet another Zen koan, then you're missing the second step. If so, no biggie. Just give yourself time to understand what I've just said.

Let's start by trying to break down these two steps into easily remembered pieces. To wit, if the patterns we're seeing are true, then it seems *hearing* involves two separate skills. We could call the first skill, the *personal auditory skill*—the ability to hear and repeat a sound. And we could call the second, the *personal intellectual skill*—the ability to simultaneously know what a sound means as we hear it, along with the ability to tell this meaning to another person.

Can you see how we've just created yet another pair of opposites? Clearly, these two things aren't just two steps. It seems, they're the baseline pair for the skill of hearing. Here, one involves sensing *information*—and the other the *meaning* of this information. And the thread which joins them—the thread of similarity—is the idea that we are not just imitating something we've heard. We are doing these things authentically.

So what does it mean if we're missing either skill? Should we be calling this lack of ability, *deafness*? At first glance, this idea seems to fly in the face of how we normally think of deafness. Then again, perhaps we can remedy this confusion by doing another thought experiment?

Try this.

Imagine you are entering the lobby of a tall building in New York City. Now imagine—as the door closes behind you—a loud sound comes from somewhere above you. Do your best to put yourself into this scene and pay close attention to what's going through your head.

Pay special attention to what you're picturing.

When I do this thought experiment, I picture myself trying to figure out what that loud sound was. What caused that sound? What did it mean? What just happened? Know I felt the same way when I first tried to hear the Greek letter-sound, ψ (psi). To me, it had no meaning. So it still just sounded like noise.

This reminds me of another story. This one took place at my dentist's office. At the time, I was in my late twenties and had gone there to pay a bill. An attractive girl I had not seen before was behind the counter. And as I chatted her up, I noticed the face of her watch had an unusual pattern—it looked like paisley swirls. Me being me—perpetually inquisitive—I asked what her watch's face meant. Specifically, I asked what the swirl-like characters on her watch were. To which she replied—with a serious face—"putzls." At which point, I asked her who told her that—and when she told me a coworker, I changed the subject.

To her, the characters on the face of her watch were called, *putzls*. To me, I had no clue. But the word-sound she used, *putzls*, did have a meaning to me. My Jewish friends had taught me that in Yiddish, a putz is a derogatory reference to a penis—as in a person is a prick. Moreover, they also taught me that Yiddish uses several suffixes for sizes. Adding the sound "aleh" to the end of a word makes a thing the child-sized version. Adding "l" or "ele" to the end of a word makes it smaller than normal.

A putz is a normal size penis.

A putzl is a small penis.

A coworker had told her the characters on her watch were putzls. They then failed to clue her in on the joke. I live in a very Jewish area. I can only imagine how many people she told this to before someone made the connection. In fact, I wince when I picture her finally being told what this sound really means. So while she and I both could hear the word-sound fine, she had no sense of the real meaning.

In effect, she'd been deaf to what she'd been saying.

Are We Ready to Define *Hearing* and *Deafness*?

Are you beginning to see how word-sounds and word meanings are separate parts of learning to hear? More important, can you see why I'm asking if sounds with no meanings equate to a kind of deafness? The thing is, we need to move on at this point, as we've asked enough questions to fill a whole book. Perhaps we'll revisit this topic in step three, when we experiment. We'll see.

For now, what's important to take away is how all this enables us to define hearing and deafness. Here, we could say that hearing is *being able to consciously witness a sound to which you've assigned a personal meaning*. This makes deafness *being unable to consciously witness a sound OR unable to assign a personal meaning to a sound—or both*.

If these things are true though, then they raise a slew of new questions. The most obvious question of all? Can seeing sign language be *hearing*? In other words, can the out-in-the-world, sound part of hearing be in some ways interchangeable with the inside-you, mental sound part of hearing? Moreover, can this be especially true when it comes to using all languages?

For that matter, should seeing signs and gestures in general be considered a kind of hearing? If yes, then when we see someone using ASL, why do we see this as evidence this person is—or is talking to—someone who is disabled? Can it be, this person is merely different from the rest of us? Can this be why many experts on deafness see deafness as difference, not disability?

Are you beginning to see the potential in a scientific method which makes you ask these kinds of empathetic questions? Isn't this how a scientific method should treat the people and things in our world?

What About the Medical Condition, "Word Deafness?"

As I was gathering data about the part *meaning* plays in hearing and deafness, I stumbled onto a phrase which connects sounds-without-meanings to deafness. Here, I'm referring to the medical condition doctors call, "word deafness." What is *word deafness*?

One source defines this condition as: the inability to comprehend the *meanings* of words—*though they are heard*—caused by lesions in the auditory center of the brain. Another defines it as: the loss of ability to *understand* spoken words, especially as the result of a cerebral lesion—also called *auditory aphasia*. A third defines it as: an aphasia in which the *meaning* of ordinary spoken words becomes *incomprehensible*.

In all three cases, a definition for a kind of deafness refers to *words with no meanings*. It seems then that I am not the first to refer to words without meanings as *deafness*. So why have I not found a single reference to *meaning* being a necessary aspect of *hearing*. Truthfully, I haven't a clue.

This points to a more significant problem with the current scientific method—how it tends to fragment scientific concepts. For instance, with regard to deafness, the current method has yet to constellate *deafness* the condition with *deafness* in general. At the very least, this prevents scientists from properly defining their terms.

In effect, *scientific fragmenting* (the use of unconstellated data) causes scientists to struggle like the proverbial four blind men. Here each scientist sees only part of the elephant. Then, no matter how sincerely this scientist tries to know the nature of this elephant, at best, he or she only knows a small portion. At worst, he or she never realizes it's an elephant.

This is why constellated science sees scientific fragmenting as something which cripples scientific advancement. To constellated science, all terms must fit all data—and vice versa. So yes, in this chapter on deafness and hearing, I've now spent more time defining deafness than on describing the ear. At the same time, because I have, I feel far more aware and curious than when I began this chapter. In fact, I feel so curious right now that I feel urges to write a whole book on deafness and hearing.

But I won't. I can't. I really shouldn't. Oh, boy. Not again.

Imagine feeling like this about science?

What About Mental Sounds You Can't Speak?

As I look back at those "word deafness" definitions—and specifically, at the word *aphasia*—I'm reminded of when I was a young intern. That year, I facilitated a stroke group which left me with a profound empathy for people who suffer mental losses. Much of this empathy came from seeing the pain which strokes inflict on peoples' families.

For example, in one case, I was told it took a man two years to learn to speak his son's names. All the while, he knew and could hear the sounds of their names in his head.

Now think back to my diagrams on the ear. Can you see where the problem was? Somewhere in his chain of conversions. In effect this man could not convert the sounds in his head to sounds out in the world. And if you're at all like me, you're now wondering where in the chain these conversions failed.

Back then, when I was an intern, I didn't know how to ask questions like this. I simply accepted the medical model's description of his condition—a stroke physiologically broke part of his brain, and that brokenness prevented him speaking words.

Today, that kind of an explanation would never be enough. Indeed, I find myself wishing right now that I'd known what I know today back then.

Know the reason this man came to mind is that his medical condition is also called, *aphasia*. This refers to a condition which occurs when a stroke affects the left side of the brain. In many cases, this condition leads to a language impairment which makes it difficult to use language. And if you look closely, the main impairment is always some kind of disconnect between sounds and meanings.

More important, were you were to read descriptions of this condition, nowhere would you find references to the word *deafness*. Again, this happens mainly because of the way science fails to connect concepts.

Okay. I have to confess, I'm tempted right now to describe all the aphasias—Wernicke's, Broca's, Global, and so on. Indeed, I'm feeling so curious right now that I'm having trouble restraining myself.

Again. I won't. I can't. I really shouldn't.

Oh, boy, here I go again.

Okay now. Big breath.

Moving right along.

What About Sounds With Multiple Meanings?

Have you ever watched babies pay attention to a sound and wondered what they were thinking? Have you ever made the mistake of thinking you knew based only on their facial expressions? How about what babies are "trying to say" when they gurgle, cry, grunt, howl, or squeal? Have you ever tried to decode these things? I certainly have.

If you have, then you have a good idea of what deafness is actually like. In these situations, you are effectively deaf. Surprisingly, in all these cases, you have both auditory and visual clues. Even so, you still cannot know for sure what these babies are thinking or expressing.

The point is, most sounds—in and of themselves—have no inherent meanings. Moreover, according to what we've found so far, before we can grasp a meaning, we must first connect this meaning to a sensory experience. This makes *information* and *meaning* two separate—but equal—partners in any and all personal learning experiences. And these experiences include learning to hear.

Know that in a future chapter, we'll explore the implications of this idea—that *information* and *meaning* are the two sides of the coin of human consciousness. There we'll look at a set of four simple, algebraic formulas which quantify and qualify certain aspects of human consciousness. These formulas give us—for the first time—the ability to scientifically and objectively represent human emotions.

Can you imagine? A scientifically sound way to quantify and qualify feelings? No more 1 to 10 scales wherein people's answers rarely go beyond guesses. This said, these ideas raise even more questions. For instance, can we be deaf to an emotion or an idea? And is this beginning to give you a sense of how complex hearing and deafness actually is?

In a way then, we could say that *hearing* is simply *the intelligent experiencing of an open mind*. Admittedly, in saying this, I'm straying a bit too far from the pragmatic side of science. The more important question to ask is, how do we decide what a sound means? And if we limit our question to language, it appears we mainly watch and learn.

The thing is, in order to accomplish this learning, we must personally connect a meaning to this sound. Here a good example would be any two year old who has just learned the meaning of a word. For example, take one of my favorite reference experiences for how the human mind learns. This experience occurred at a birthday party for a one year old.

As I think back to this event, I can picture myself standing on one side of a table. In front of me is a highchair containing the birthday boy. Standing next to me is the father of a two and a half year old boy and he's holding his son in his arms. And as I stood there, I noticed that the one year old birthday boy had a circle of snapped-together, plastic blocks on his wrist.

At some point, the two year old tapped my shoulder to get my attention. He then pointed to the baby bracelet and said, "baby watch." Sadly, no one—including this boy's father—recognized the significance of what this boy had just done.

His information?

The snapped-together baby blocks the one year old was wearing on his wrist.

He then became a little scientist and explored this information. And at this point, had recognized two patterns—the pattern of a baby, and the pattern of a watch.

He then combined these two patterns and created a class of objects we adults do not have.

He created a class of objects called, "baby watches."

Of course, when this two year old later learned the more universally "approved" meaning of the word *watch*, he ceased to be able to make these kinds of personally creative connections. Which makes me wonder how we adults decide on a meaning, when a sound has more than one.

For instance, consider homonyms—words which share a sound but have different meanings (e.g. bass / base, cell / sell, browse / brows). These sounds can—and sometimes do—create auditory confusion (what does this sound mean?). In part, this confusion comes from something deaf people and folks with Asperger's share. They tend to see the first meaning they learn for a word-sound as the one and only correct meaning. They then have a hard time assigning a second meaning to this sound.

This multiple meanings thing also explains why folks with Asperger's notoriously fail to grasp jokes. Here, the problem is that many jokes rely on alternate meanings for the same word-sound. Sexual jokes are especially prone to this kind of auditory double talk. In fact, I've been told Shakespeare filled his plays with this kind of thing.

So is the inability to assign alternate meanings to word-sounds yet another kind of deafness? Can being unable to get a joke be considered a kind of deafness as well? Indeed, can being unable to understand Shakespeare also be a kind of deafness?

It seems the new method is currently working rather well.

The Brain's Electrical to Chemical "Medium Conversions"
(The medium changes. The pattern stays the same.)

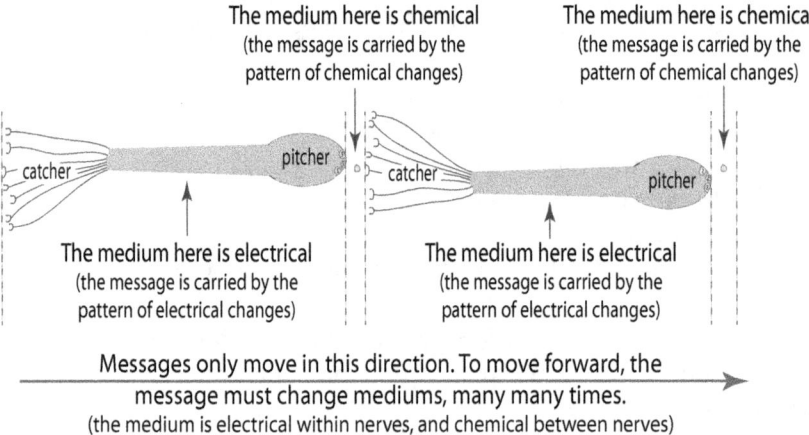

What About the Brain's Internal Conversions?

At this point, a few people may feel the need to point out what seem to be inaccuracies in my drawings. Here, I'm referring to my references to electrical patterns, when in truth they are electrical to chemical patterns. Even more accurate would be to say they are chains of electrical to chemical conversions. Here again, the word to focus on is, "conversions."

Indeed, in a chapter on deafness, the idea of *conversions* may be the most important concept of all. Each of our senses, including hearing, relies entirely on a sequence of conversions across mediums. Indeed, these conversions across different mediums are what connect us to our world—and our world to us. And the mediums—the air, skin, bone, nerves, or neuro-chemicals—simply act as carriers for the underlying patterns.

Realize, it's the underlying patterns—and not the mediums—which communicate the actual messages. Please keep this in mind in the next section, when we discuss where our sense of reality comes from. There we'll explore the part our five senses play in creating our sense of reality. But first, we've got a few things to discuss, including how media conversions underlie radio broadcasts. Why? Because if you grasp how radio broadcasts happen, the rest of this chapter will fall into place.

Take your time. And bon voyage.

FM Radio - The Changing Frequency Carries the Message
(The medium changes. The pattern stays the same.)

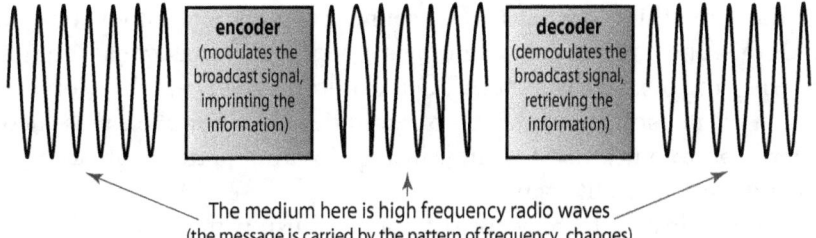

The medium here is high frequency radio waves
(the message is carried by the pattern of frequency changes)

FM Radio Broadcasts as Medium Conversions

Anyone who knows me knows how I love to find parallels to human nature in technology. These parallels have much to teach us about who we are, since we create our technology in our likeness and image. This likeness and image includes the way radio broadcasts work. And here, what I'm referring to is the concept of "carrier waves."

What I mean is, radio broadcasts use carrier waves to "carry" their messages to you. And all of us know that radio stations broadcast on different "channels." We also know we refer to these channels by their broadcast frequency. Thus channels are named things like 98.6 and 106.7.

In addition, we know that to listen to a station, we must tune in to this station's frequency. But what is a "frequency," and what are we tuning in to?

The "frequency" of a radio station is the number of times its carrier waves "wave" in one second. And what makes it a "radio" wave (as opposed to a *micro*wave) is how *frequently* it waves in that second. Radio waves wave between 88 and 108 million times per second. Microwaves wave between 1 and 300 billion times per second.

But what is a radio wave anyway? It's a wave in an electromagnetic medium. And radio stations make these waves in the electromagnetic part of our world like audio speakers make waves in our air. With audio speakers, a physical cone vibrates in and out, and as it does it create waves in the air. Indeed, were you to draw a speaker's movements as seen from the side, you'd find they much resemble the radio waves in the drawing above.

What about the term, FM, as in an FM radio station? FM means "frequency modulation," a term for how this medium—FM radio carrier waves—carry information. In the case of FM carrier waves, something "*m*odulates" (changes) the carrier wave's *f*requency slightly. These frequency modulations (slight variations) literally imprint the carrier wave with analogous patterns of physical, audio information. Again, it's just changes in mediums, this time from air to electromagnetism and back.

The same thing happens when sound waves vibrate (modulate) our ear drums. These patterns of vibrating sound imprint analogous vibrating patterns onto our ear drum heads. Our ear drums then pass this vibrating pattern on to the three bones in our middle ears. They then pass these patterns on to our inner ears, imprinting vibrating patterns of electrical change onto our auditory nerves.

Are you lost as to how radio stations transmit messages? If so, try picturing this. Imagine you're on a beach and that you're watching waves crash on the shore. The "frequency" of these waves crashing onto the shore is the number of times they roll in and out in a second. And according to scientists, a common frequency for ocean waves is roughly 1.8 waves per minute. Which means a frequency of about .02 to .04 waves per second.

Now recall the frequency of FM radio carrier waves. They "roll in and out" between 88 and 108 million times per second. So how does the information get imprinted on the medium? To see, picture a chain of signal drummers sitting on a series of mountain tops. Now think of how these signal drums send signals. First, a drummer's hand hits a drum head. This causes this drum head to vibrate at a certain frequency. This vibrating head then vibrates air, and vibrating air IS sound. So whether the drummer hits a single hit or three in a row, the frequency (the musical note) of this drum sound will stay the same.

What about the actual messages? The drummer uses agreed upon patterns of drum hits to convert and transmit these messages. These changing patterns of drum hits literally *encode* spoken language into drum hits. Then, when this pattern gets "received" (heard) by the final drummer, this drummer *decodes* this pattern. By this, I mean this last drummer converts this pattern of drum hits back into spoken language.

Radio stations act like the first drummer. Car radios act like the last. What I mean is, radio stations modulate the carrier wave like a drummer hits (modulates) a drum head. In doing so, the radio station encodes message patterns onto these waves. Your car radio then tunes in to this frequency and strips out the carrier wave. This then leaves only the message—which then gets converted back into audible sound.

In a way, this same thing happens whenever we hear sound. We take in a real world signal, then encode it on different mediums in our bodies. Here, we go from ear drums, to middle ear bones, to inner ear tubes, and so on. Finally, this signal reaches our brains, where our brains decode it.

The thing to see here is simple. First we must be able to hear this sound. Then our bodies must encode it and send this signal to our brains. Our brains then decode this signal—but only if our brains know the code. Like signal drums, the code is the meaning we attach to the sound.

[4] Deafness and Hearing: What Do Deaf People Hear?

When I first wrote this subhead back in Cartesian step one, I wrote it as this chapter's initial, fourth line of questioning. Back then, I had no clue I'd get so deeply mired in philosophical questions. I never imagined I knew so little about hearing. To be honest, I would have preferred to focus mainly on the physical, practical science of seventh grade. The problem is, we must include scientific philosophy or risk incompleteness and scientific fragmentation.

Fortunately, constellated science gives us tools to address both the philosophy and the practice. Moreover, it allows us to do this in ways which cause a synthesis of both to emerge. So has this happened yet in this chapter?

Let's see.

To begin with, we've been creating and refining our definitions throughout the data gathering process. Especially important in this section is our current working definition of the word *hearing*. After all, the question we're asking here is, what do deaf people hear? And to ask this question, we must have a real definition for hearing.

Before we turn to this definition and its implications though, I first want to acknowledge what we've accomplished so far. I also want to mention how much this process has affected me personally. For instance, were I to have asked myself—at the beginning of this chapter—what Deaf people hear, I might have answered, "next to nothing."

And now?

Let's start with the idea that during this process, we defined a baseline pair for deafness. People with Type 1 deafness hear no out-in-the-world sound. People with Type 2 deafness do. This means, if we limit our question to whether deaf people hear out-in-the-world sound, we have a clear answer. People with Type 1 deafness hear no out-in-the-world sound whatsoever. Thus they receive no out-in-the-world auditory messages. Whereas people with Type 2 deafness hear at least some out-in-the-world sound. But this sound is missing some aspects of what turns these sounds into messages.

What about what I've been calling *mental sound*? Do deaf people hear mental sound? To be honest, since we don't currently have a scientific definition for "sound," this question is hard to answer. My first instinct is to say, yes. Deaf people probably hear some version of mental sound. For instance, we know the brains of babies allot space for this processing. But exactly what this sound sounds like to these babies, to be honest, I haven't a clue.

Moreover, because current research fails to adequately define the word, *sound*, I'm not sure there's even a way deaf people can answer this question yet. At least, not scientifically. So much for these questions. As I think about it though, a better line of questioning might focus on how people with the two types of deafness experience mental sound.

To wit, from my research it seems a lot depends on when the person became deaf. Those born with Type 1 deafness don't seem to have the reference experiences needed to understand this question. Whereas people with Type 2 deafness do have a lot to say about mental sound. But what they say often leaves me feeling even more confused.

The Hearing to Speaking Feedback Loop

One clue here may lie in the way our brains use a voice-to-ear feedback loop to adjust how we speak. Granted, people with Type 2 deafness have far less information than hearing people with which to do this. This said, deaf people can—and do—make these adjustments. Thus they must hear something. And this something must include some form of mental sound.

So what do they say they hear? Not surprisingly, it varies from person to person. Moreover, it varies so widely that it almost makes no sense. Some deaf people report hearing mental sound, including word-sounds. Others report their minds focus more on the visual aspect of language. And others report having no mental sound.

At the same time, one Deaf person reported he sometimes—simultaneously—reads things, and hears a voice, and lip reads, and sees signs. Clearly this person experiences the same kind of synthesis hearing people do, only with even more data sources—and with more personal awareness. Awesome. This said, even this brief survey leaves us with a lot of questions. Here, perhaps the most important of all is, how can we better ask this question? Can we even currently ask it, given our current lack of clear definitions?

For instance, I spent many weeks collecting data about deaf people and mental sound. Not once did I find a clearly reported reference I could actually call scientific. This is not to say I found no research. To be honest, I found many focused, sincere efforts. In particular, I was touched by the book, *The World of Deaf Infants* (2004) which documented a 15 year longitudinal study. Here, I found immaculately documented, scientific efforts which yielded all sorts of useful data. However, even here, none of this work was grounded in scientifically universal definitions—not for *hearing*, nor for *deafness*, nor even for *sound* itself.

More troubling were the glaring omissions, such as the effect mental focus has on hearing. Here, I'm referring to the idea that humans have the ability to tune out sounds—to purposely or unconsciously not hear them. Moreover, as any audio engineer would tell you, this skill is at the heart of an audio engineer's trade. Without it, a sound engineer can't make useful adjustments to things like orchestras or mixed groups of talking people.

So do deaf people have this ability? Scientists say this skill is standard equipment on all human babies. Here we're talking about the ability to focus in on things while tuning the rest out. In truth, I can't imagine how we could possibly learn things without having this skill. Also, deaf people report being able to do this with vision, taste, touch, and smell. So is it possible they do it with all those things, but not with sound?

This raises the question as to how many deaf people deliberately shut out sound? Does this question seem crazy? Then research how many Deaf people are opposed to cochlear implants for this very reason. I, myself, love the sounds of pulsing music and children's laughter and babies vigorously protesting in public. But I also have a deep love for the silence of the forest, or the night-time stillness of newly fallen snow, or the presence of deep-Earth cave spaces.

I also hate dogs barking endlessly, especially when I'm trying to sleep. In those times, I can easily understand the beauty in deafness.

As for our own working definitions, allow me to repeat them here. Let's start with hearing.

What is hearing?

Hearing is *the intelligent experiencing of an open mind*, where doing this requires two skills. One is the *personal auditory skill*—the ability to hear and repeat a sound. And the other is the *personal intellectual skill*—the ability to simultaneously know what a sound means as we hear it, along with the ability to tell this meaning to another person.

In addition, we have a technical definition as well. *Hearing* is *the ability to convert, recognize, understand, communicate, and recall electrochemical analogies for sensory patterns of sound.*

What about deafness?

Deafness is *the inability to experience an open mind*, due to deficits in one or both of the skills required to hear. One class of deficits involves losses in the *personal auditory skill*—the ability to hear and repeat a sound. And the other involves losses in the *personal intellectual skill*—the ability to simultaneously know what a sound means as we hear it, along with the ability to tell this meaning to another person.

Again, we also have a technical definition as well. *Deafness* is *the inability to convert, recognize, understand, communicate, or recall electrochemical analogies for sensory patterns of sound.* Essentially, any break in the conversion chain equates to a kind of deafness. This includes semantic breaks.

Finally, what is *sound*?

Sound is any message carried by any medium wherein vibrations in a physical medium (e.g. vibrations in air, metal, water) get converted into patterns analogous to these audible changes. This includes—but is not limited to—vibrations in ear drums, vibrations in the bones of the middle ear, physical movements in the inner ear, electrochemical changes in the auditory nerves, and any mental impressions which result in patterns analogous to audible changes.

So what do deaf people hear?

According to these definitions, they hear a lot.

What Makes Learning Language Have a Season?

Let me start by saying, I'm almost seventy and have been using constellated science for years. During this time, I've used it to understand all manner of things—and to discover many things as well. So why, with this tool at my disposal, did I struggle so to learn Greek? For that matter, why did I fail to retain most of what I thought I'd learned?

Did this inability happen because I'm old—and old people struggle to learn new things? In some ways, yes, this is true. But this cannot be the whole story. My evidence? A limitation in the way our brains store data causes this same thing to happen to children who become deaf at an early age. James is a good example of this.

This makes me wonder if normal old people—and deaf young people—can be the baseline pair which could unlock language learning difficulties, and maybe even learning difficulties in general. Can this be true? To see, let's start by looking at the storage problem I mentioned a moment ago.

The story of this storage problem begins with understanding how the brains of babies store new learning. Admittedly, what I'm about to say is yet another gross oversimplification. This said, it begins with the idea that as babies, we store things pretty evenly across our brains. Then, as we learn and grow—and as our senses differentiate—our brains begin to cluster information in specific places.

This is not to say our sense of hearing—or anything else—is stored all in one place. But things are stored in clusters which make them vulnerable to losses. For instance, if the part of your brain called Broca's area suffers damage, then your ability to speak and write will suffer too. At the same time, you'll still be able to read and understand spoken language. But if the part of your brain called Wernicke's area gets damaged, the opposite will happen. Here, people tend to [1] speak in long sentences which have no meaning, [2] add unnecessary words to what they're saying, and at times, [3] create new words. So while these folks can still make normal speech sounds, they have difficulty understanding the meaning of speech and a hard time realizing their mistakes.

Now constellate this pair of opposites, then consider what this tells us about how the brain works.

Obviously, since babies can't read, write, or speak, if this happened, you wouldn't see much evidence for this damage. Over time though, most babies can and do learn language. As they do, their brains store language-related, how-to instructions in these two areas. And gradually, these two areas differentiate more and more.

But what if something—such as deafness—impairs a baby's developing sense of language? If this happens, then this baby's brain has little to nothing to store in these two areas. Moreover, because the brain abhors unused space, it allots the unused space to other developing senses. Indeed, this reallocation even has a name. It's called, "cross-modal, cortical reorganization," and it can happen even later in life.

In a way then, what I'm describing is similar to what might happen if you arrived at a wedding reception and saw an open seat at a better table. You might walk over and see a card which says this seat is reserved for someone else. After a while though, if this someone never arrived, you might take this seat. And for a while, this might even go well.

Now imagine what you'd feel like were this person to eventually arrive—confused, at the very least. Imagine also that you're halfway through dinner when this person arrived. Obviously, moving now would not be so easy. Nor would you likely want to move.

Now consider what happened to James's brain. James lost his hearing at six months old. He then got cochlear implants at two. In between these times, his brain assumed language was never coming. And if we apply my wedding guest analogy to James, we'd equate the unused places in James's brain to the open seats at the wedding. Then we'd then equate language to the late arriving guest.

Imagine how confused James must have felt after getting cochlear implants at two?

In effect then, our DNA creates the equivalent of seating-arrangement cards in our brains. Here, our brains expect certain things to sit in certain places. If these things never arrive, then our brains scramble to rearrange the seating. After all, we've paid for these seats, whether empty or not.

Brains usually resolve this problem by seating other guests in the empty seats. And if the yet-to-arrive guests never arrive, this works out fine. But if the original guests show up late—such as when James got implants at two—there is no easy way to undo what has already happened.

Something similar happened to me when I tried to learn Greek. My brain's language storage space was designed to hold the patterns of English grammar and syntax. Conversely, the opposite thing happens when a young person grows up hearing two languages. At a young age, the brain is still designing the language storage space. So the person's brain designs these storage spaces to accommodate both language's patterns. In effect, it allots more seats at the tables.

Admittedly, there's a lot more to how the brain allocates space. In fact, at the risk of getting you mad, I have a confession to make. Recently, I made what I consider to be one of the more important discoveries of my life. This discovery involves what we've been talking about—how the brain allocates storage.

In truth, it appears this discovery may solve many mysteries about the brain. This includes the two mysteries we've been talking about here—why James never learned to speak even though sound reached his brain, and why I retained next to no Greek, despite my having a genius IQ and an amazing new scientific method. It also appears to offer solid, scientific explanations for things as diverse as what minority personalities like Asperger's and ADD actually are—as well as why most adults struggle to learn new subjects, what makes healing things like addictions so difficult, and what makes the physiological brains of geniuses so prone to discoveries. More on this in a moment.

The Brain's "Baby Shape" Game

the language container

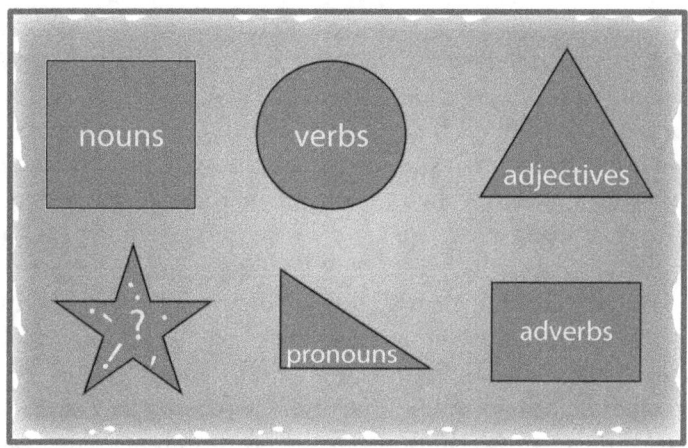

the language content

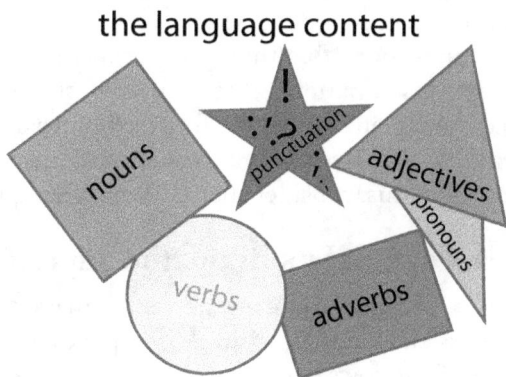

Do Baby Brains Begin to Learn Right Away?

Before we move on, I need to mention one more idea. This idea is that, while it's our DNA which makes out the seating cards, our brains also get a say. Admittedly, the brain's say is more like the groom's when it comes to wedding planning. Here, the brain's organizational input gets considered only after the DNA has spoken.

Translation? It's our DNA which lays out our brain's basic floor plan. This is why we all have roughly the same general layout for where things get stored. However, before the brain can store content, it must first design and create storage containers. Here, our DNA tells the brain where to build these storage containers. But our brains design the way the space gets used inside these storage containers.

Where do these designs come from? To begin with, the human brain is pattern based. IQ is pattern recognition, remember? This is why the logical geometry underlying the new scientific method works so well. Logical geometry is like a child's block-in-the-hole game, with six holes in the board. It seems our brain comes with these six patterns installed, albeit, we never realize that they are there. But we use them nonetheless, first to look for patterns—then to constellate these patterns into groups.

Know that when I say we look for patterns—and ways to group patterns, I'm describing the essence of learning itself. These patterns and groups of patterns ARE what we learn. Moreover, because the meaning of things comes from the connections they make to each other, each time we discover new connections, our lives gain meaning.

For example, imagine the aha you felt when you first discovered birthday candles. And lightning bugs. And flashlights. And campfires. Now imagine the grand aha you felt when you first constellated these things and realized, they all light up spaces.

All learning centers on sifting through life experiences while looking for patterns and groups of patterns. The brain then stores and uses these patterns to process and plan our lives. Before the brain can do this though, it first must create a warehouse to organize and store these patterns. And to see what I mean by this, consider the following thought experiment.

The Early Years of a Baby's Brain: the Supermarket Story

Imagine you've decided to open a new supermarket. This morning, a real estate agent is showing you a space. You park and as you walk up, you see a big wall of windows and doors. You also notice the parking lot is large and there's good access from the street.

Now as you walk in, you see a huge empty space. This space has all basics—white walls, lights, heat, air conditioning, and bathrooms. And of course, you see the ubiquitous floor to ceiling steel support columns. Obviously, other than these basics though, there would be no place to put the food. No aisles of shelves. No candy racks. No cold storage. No veggie bins.

So now, let me ask you a question. If, at this point, you signed the lease, would you be ready to take delivery on the food?

Take a moment to try to picture this question.

Of course, the answer is no. Where would you put everything? In one big heap in the middle of the nice new floor? What about the frozen food? How long would it last? Or the vegetables, or bananas, or cherries? Moreover, how would people find what they want? By picking through the pile like Black Friday shoppers? Obviously, having the food delivered would just create chaos.

Oddly, baby brains face the same dilemma. They arrive with all the basics—heat, lights, air conditioning, ceiling supports, and so on. They also have a great big space in which to store all the baby's life experiences. Similar to the brand new supermarket space, however, a baby brain isn't ready to store things. Where would they go? In a great big heap in the middle of this brain? How would the baby find anything?

Curiously, I cannot find a single researcher who poses this question. So far, it seems we all assume baby brains begin to store memories right away. In one big heap. In the middle of the floor. In effect, we assume, babies begin to learn from day one. And we assume this learning amounts to acquiring knowledge and piling it in heaps.

Supermarket Design in Four Easy Steps

My point? To see it, we'll need to explore this idea a bit further. When we last spoke, you'd signed the lease on a big new, open space. Now you need to figure out where everything will go. So okay. This wall? Frozen peas and spinach—cold storage with doors. And that floor space? Potatoes, onions, and mangoes—open vegetable and fruit bins. And so on. And so forth. Here's the hard part, though, and please, go slow and really try to take this in. Supermarket buildings are not *food*. Nor are supermarket shelves, *food*. Supermarket buildings are storage *spaces*. And supermarket shelves are storage *containers*.

In truth, were you to actually want to open a supermarket, we could break this down into four steps (no surprise here). Step [1] acquire a big empty space. Step [2] walk around this space while doing thought experiments as to where shelves, racks, bins, and freezers (the storage *containers*) would go. Step [3] design and build these shelves, racks, bins, and freezers (the storage *containers*). And Step [4] stock these shelves, racks, bins, and freezers with food (the stored *content*).

Now let me ask you something else. Could you skip any of these steps? For example, could you skip Step [2]; doing the thought experiments and designs, and Step [3]; building the storage containers? In truth, you could physically do this. But you'd likely end up with a heck of a mess. Indeed, you might even end up throwing most of the food out.

Know that baby brains must go through the same sequence of four steps. Step [1] happens when they get born. This is when they acquire the space. Step [2] begins a moment later, when the thought experiments begin. But at this point, they aren't ready to take Step [3]. Why? Because designing storage containers requires you know what you're storing. So during Step [2] and before Step [3], baby brains have no storage containers.

What does this time look like? To see, watch any normal six month old. Each time this baby picks something up, you'll see him or her looking to discover patterns. What comes next though? Just one thing—he or she quickly loses interest. Then the baby moves on to the next thing. And so on, and so on—day after day.

Can you see how this explains why most of us cannot recall events from this time in our lives? Contrary to what we've assumed—that our brains store everything we experience—until we build places to store these experiences, we pick through them then toss them out. So am I saying we don't remember these early experiences because we did not store them?

In truth, this is exactly what I am saying. And this happens to be one of the more important things to know about baby brains.

In their first years of life, baby brains mainly work on Step [2] and Step [3]. In other words, baby brains seek and constellate patterns of life experience. Then they use what they find to design and build storage spaces which will hold baby life experiences.

Realize too that during this time, baby brains are still under construction. This means they can't open for the business of learning until the basic storage containers are done. In effect, Step [3]'s arrangement of storage containers functions like five subject, high school notebooks. Without them, it's hard as hell for babies to take notes.

Creating Storage Spaces with Differently Shaped Holes

Admittedly, what I've been saying is a bit abstract. So perhaps it might help if I offered a few real life examples. Let's start with how we learn language. To begin with, our DNA allots language its own storage area within our brains. Then, as we're exposed to the sights and sounds of language, our brains identify the patterns in this language.

Then they build cubby holes for each and every one of these patterns. At the same time, our brains do not build cubby holes for the patterns we are not exposed to. Know this idea is important. It accounts for why babies learn so quickly. It also accounts for why older folks struggle to learn new things.

New things require new storage containers—containers designed to hold the patterns in this new learning. Once baby brains build storage containers in each of the allotted spaces though, they cannot simply build new containers. Nor can they alter existing containers. Why not? To see, let's return once more to the supermarket metaphor.

To Remodel a Supermarket, Must You Close It?

Okay, say that at this point, it's been many years since you opened your supermarket. And in some ways, what was once exciting has become flat, boring, and old. So you decide you want a make-over—a complete redesign; a brand new look. And you hire a designer and she introduces you to some completely new ideas.

The thing is, you can't simply move things around and accomplish what you want. And while many things will roughly stay the same, some things must completely go. For instance, say you want to add a self-serve salad bar—a complete make-your-own lunch counter. Something would have to be made smaller to make room, meaning, something would have to go.

Now consider how different this would be from merely refining the supermarket's existing storage. With the existing storage, you might be able to move shelves around. You could also definitely rearrange where different foods go.

Know adult brains do the same thing. They keep refining what we know all through our lives. When it comes to learning unfamiliar patterns though—such as the letter-sounds in a strange language—this doesn't work. We have no storage container for these patterns. So most of what we see and hear simply passes right through us. And yes, if we make the effort, we can temporarily remember some things. But permanently store them? Without a storage container, we cannot do it.

How then can some people learn new languages later in life? To be honest, as usual, I only have questions like could something have happened in these people's lives—something akin to supermarket remodeling? For instance, could they have moved to a foreign country where no one speaks their language? And could this have caused them to experience what baby brains experience?

Then too, think about it. To baby brains, getting born is a lot like moving to a foreign country. Can this be why immersion is said to be the best way to learn a new language? Can this also be why—when most of us try to learn a new language later in life—much of what we hear goes in one ear and out the other? No storage container—no real memory?

Containers, Content, and Storage Spaces

Can you see yet how what I'm suggesting here all boils down to one idea? The idea? That all learning could be divided into two categories. Category one? Baby brains learn how to design and build storage *containers*. Category two? Baby brains learn how to store *content* in these containers.

It seems baby brains learn to do both of these things. They also seem to learn them in this order. And if no storage container gets created for a particular group of patterns? Then these life experiences get tossed.

However, if the baby can connect the new group of life experiences to the patterns already in a container, then it refines this already existing space, and the life experience gets remembered.

Is what I'm saying still confusing? Then consider the drawing—the baby "shaped blocks into shaped holes" game—you saw a few pages back. Say the blocks are "the patterns of your native language" and the shape board, this language's storage container. Discovering each of these patterns would be what you do in Step [2]. Using these patterns to design and build the board would be what you do in Step [3].

Now consider how different language container boards would use differently shaped, "pattern holes." For instance, most of the patterns for English don't exist in Japanese. In Japanese, there are no "L" or "R" sounds. But there is a letter-sound in between L and R. Which is why, when Japanese adults try to learn English, their brains simply have no place to store—or even process—these English language "shapes."

Force-Fitting Shaped Blocks into Already Existing Holes

Now consider what happens when you hear someone speak a foreign language. Your brain contains a native language container with a limited number of letter-sound patterns. Moreover, since all learning is pattern based, we first try to fit what we hear into the closest letter-sound holes in our brains. This in fact is what we call, translation.

But if no such hole exists? Then at best, we simply hear noise. Worse yet, consider what happens when a native Japanese person hears an English "L" or "R." Because their language contains a pattern roughly in the middle of these two sounds, they feel compelled to force fit these two sounds into that particular Japanese language sound hole.

Unfortunately, this hole has no English equivalent. So to us, Japanese folks simply reverse these sounds. Sometimes. In truth, they literally cannot hear these English sounds.

No such patterns exist in their native language containers.

Are Baby Brains Like Computers?

Is this still not making sense? Then consider how computers are like brains. Here, computer operating systems are the containers, and the data is like baby life experiences. Unlike baby brains though, all computer operating systems are built like onions. They all employ nested layers (which are storage containers) where each layer stores one kind of data.

Most people have even heard of at least one of these layers—the innermost one, the one programmers call, the "kernel." There's also a layer which connects the kernel to the hardware (keyboard, monitor, etc.), and another which connects the hardware to the programs. Admittedly, this description is way oversimplified. But my point is, all computer operating systems employ *containers* (the nested layers in their operating systems, for instance), within which they store their *content* (data).

Why tell you this? Because this is what I drew on when I created my personality theory. Human brains design computer operating systems. So I assumed that reverse engineering these computer operating systems would tell me a lot about human brains. It turns out, doing this revealed what had been missing from all other theories of personality. All other theories fail to separate learning into patterns for containers and the patterns of the content.

Ironically, I referred to this idea (containers vs content) in detail throughout Book II. Yet until I wrote this chapter, I never realized how this concept applies to the way all brains learn. I've struggled for years to understand what kept James from learning language—this despite his getting cochlear implants which allowed his brain to respond to sound. And I've also struggled to understand why, despite my IQ, I failed to be able to learn Greek.

Did James and I both suffer from the same problem? Was our problem that we couldn't find a way to close our supermarkets for renovation?

What Makes My Theories So Hard to Learn?

At this point, I have a confession to make as to the main reason I've been driven to solve this problem. Know I've spent more than two decades trying to pass on my discoveries to others. In all this time, I've repeatedly failed to find a way to do this. It's seemed, there is no way to do this.

For example, in my practice as a therapist, I've helped hundreds of people to positively alter their natures. I'm not bragging. They tell me this all the time. Yet when people ask my clients what it is they do in my office, they all tell me they don't know what to say. Nothing comes to them.

Now consider what I've just said. How is this possible? My therapy IS to teach people to recognize the patterns in human nature. This includes teaching them to recognize the patterns in their own natures. And because you can't change what you can't see—I repeatedly tell them what I'm doing.

Despite these constant attempts to teach them, by connecting these ideas to real life experiences, these folks cannot describe what it is we do. And to be honest, this has hurt me. I've felt like a failure. Until I wrote this chapter and finally realized the problem.

Separating containers from content IS a pattern. It also seems it may be the foundation concept underlying all learning. Obviously, neuroresearchers do not get taught this concept. Nor do educators, nor therapists, nor anyone other than my clients.

This means, people experience this pattern like fish experience water. Then like native Japanese speakers trying to hear "L" and "R" sounds, they try to force fit this idea into the "learning holes" they already know.

Where's my proof this pattern exists though? Even casually observing how babies learn reveals much evidence for it. Anyone who watches babies putting shaped blocks into shaped holes can see they spend most of their early years researching and creating ways to organize and store content. They also often try to force fit shapes into holes which are not exactly correct. At least until their brains create a pattern in their brains' "shape" storage containers, at which point, this game becomes too babyish for them.

Then there's the idea that almost everything nowadays has an operating system—moreover, that all these operating systems are built in layers. Computers. Smart phones. Televisions. Car engine "brains." They all separate containers (layers) from content (data). This in fact is the basic idea behind geometry itself. The shapes are the containers which organize our lives. What goes into these containers is us and our things.

Despite this "containers and content" pattern being a part of all things in our world, somehow, we overlook its importance. And of course, telling you it exists does not solve the problem. If you have no place to store this idea, it just gets thrown out.

So can being in the wordless state of baby minds be the only way we can create new containers? And is this what happens to people when they experience a significant trauma? And what about people who spend years in a wordless state such as meditation? I myself spent the ten years prior to my first discovery in this state while I explored my own suffering. Can I have built a new container for personality in this time?

Is this also why feral children never learn to talk, why children born deaf never develop the grammar and syntax of hearing people, and why

James found it so hard to say a single word? In all these cases, can the problem be that they never developed a spoken language container? In James' case, his cochlear implants sent sound-based conversions to his brain. But once the container space got allotted, is this phase pretty much done?

The bigger question is, is there a way to overcome this problem? Indeed, this may be this chapter's most important new line of questioning. Obviously, this topic could also easily take up a whole book. And I've already far exceeded the space my brain allotted for this chapter—I planned on making it ten pages. Ha! The best laid plans and all that.

Before we move on to Cartesian step three though, know I plan to explore this topic in depth in a future chapter.

Cartesian Process ~ Step Three: Experimenting

In step three—experimenting—we'll be looking to discover new facts. Specifically, we'll be looking to uncover things which make us even more curious. Remember, to be a fact, a thing must reference three qualities—a specific time, a specific place, and a specific observable event. Omit any of these qualities and this thing is not a fact.

At the same time, all experiments must also generalize to ideas. Experiments create meaningful sequences of facts, and ideas summarize groups of experiments. Keep this in mind then—that to be an experiment, this experiment must come from an idea and result in facts. So while your ideas may be grounded in logic and science, even well grounded ideas are not facts.

Here's my first example. I'll begin with ideas, then offer a fact or facts.

What's the Worst Sickness a Human Can Get?

The following story involves the ear and deafness—but not the nature of hearing per se. Rather, it's about an experiment I did which made me question the entire nature of scientific reality. What is reality anyway, and what is science looking to know? Does science want to know what literally exists out in the world? Or is science content to use the logical / experiential sum of all of our senses to create mental models?

The event which inadvertently triggered this experiment occurred at the end of an enjoyable New Years Eve. Even now, I can picture myself getting into bed that night, then letting myself fall backwards onto the pillow. In that moment, I felt happy and content, and I was simply clowning around. But my happiness didn't last long. By morning, my condition had evolved into one of the worst sicknesses of my life. It also led to what may have been the most difficult experiments I've ever done.

The actual experiments began that morning, when I woke and turned my head. As I did, it felt like the whole world had disappeared. What I mean is, as I rolled to my side, I felt like I'd fallen into a bottomless pit. And when I tried to catch myself—by pushing my hands back into the bed—it quickly got worse. The harder I pushed back into the bed, the more I felt like there was no bed. Nor hands. Nor anything else physically solid. I literally could find nothing to hold onto.

I later learned this sickening experience has a name. It's called, *subjective vertigo*. When people get it they feel as if their bodies are spinning and falling—even when they hold perfectly still. Contrast this to its complementary opposite, *objective vertigo*. When people get this, they experience themselves as still. But the stationary objects around them move—the classic "I drank too much and have the spins" kind of vertigo.

Does *subjective vertigo* sound manageable? Believe me, it's not. Most people throw up constantly and end up in a hospital on serious drugs. The thing is, many years prior, I had taken a vow to not suffer without gain. So I forced myself to begin to explore my sickness.

As I lay there in bed then, I began to experiment. I wanted to see what made my symptoms worsen. What I found was, if I was on my back and rolled my head to the left, four seconds later, the sickening falling-feeling sensation would begin. However, if I then held my head still—and waited nine or ten seconds more—the vertigo would disappear.

Okay. Great, I thought. I can get it to stop. Don't roll or turn my head to the left. Like I could do that for the rest of my life. Ha. Good luck. I soon learned that when I rolled or turned my head in the other direction, this movement would trigger it as well. So much for my plan.

In effect, regardless of the direction I rolled or turned my head in, if I was on my back then turned, I'd lose touch with the physical world. If I then tried to find something solid, I felt like I was pushing my hands through air. Overall, I felt as if I was endlessly falling into a bottomless, black pit. And all I could manage to do during this fall was to freeze and pray for it to stop.

Curiously, I realized—while falling—I also lost my ability to hear. This made me retreat even further into myself and compelled me to close my eyes. I'd then try to remain perfectly still, hoping the spinning would end. But each time the vertigo returned, it felt like it would last forever.

Of course, like most people I know, at this point, I self diagnosed. It had to be a fatal brain tumor. What else could it be? When I called my doctor, though, and described my symptoms, he calmly told me the medical diagnosis; *benign positional vertigo*. Kind soul that he is *not*, he then added that while I wasn't going to die, I might wish for death for a few days.

Thanks a lot, Doc.

What was wrong with me?

Do you remember the gyroscopes we have in our inner ears? As we age, crystallized debris accumulates on the inner walls of each of these six fluid-filled tubes. Unfortunately, if you jolt your head in just the right way, a particle of this debris can break free and start floating around. After that, anytime this particle floats through the fluid, your brain thinks your body is moving.

The problem of course is that the brain cannot make sense of the way this particle is moving. None of these patterns resembles what your brain has ever seen before. In other words, the brain has no container for this kind of motion. Thus it cannot understand it. So the brain being the brain, it then tries to guess what's going on.

Unfortunately, when multiple senses contradict prior experience, it's mostly a crap shoot as to how this turns out. Me? There wasn't much that wasn't contradictory. Eyesight? Hearing? Balance? Touch? It all failed to make sense. Can you imagine losing all your senses at once?

Is Reality, "Literal," or Do Our Brains Create Reality?

Now please forgive my struggles as I try to explain what I learned from this. Let me start by saying what I learned left me confused and even terrified for a while. What confused and terrified me was the realization that what we see, hear, smell, and touch doesn't exist. At least, not literally. In other words, it doesn't at all exist in the way we believe it does.

In truth, our brains treat our sensations as if they are an artist's water colors. The brain then plays artist and blends these sensations into something we can understand—our picture of reality. The thing is, by the time our brains create this picture, what we've sensed has gone through many conversions. This means, we never actually, directly connect with our world. Thus we cannot know what the world is really like. We can only know the picture of reality our brains create.

To be honest, I'd heard people say things like this many times before. But until I got Benign Positional Vertigo, it all just felt hypothetical. Moreover, I'd thought, what difference does it make? I'm seeing, hearing, and so on, fine. So what difference does it make? To see, consider what it's like to have Benign Positional Vertigo.

To begin with, Benign Positional Vertigo is a rather strange condition. The entire condition is caused by a single errant, physical particle. Yet whenever this particle moves, it destroys our brain's ability to construct our reality. Why? Because, like constellated science, in order to create our sense of reality, our brains must follow certain rules. *And these rules happen to be the very same rules constellated science follows.* More on this later.

Is the Brain the Artist of Our Sense of Reality?

Everything I realized came from trying to make sense of my condition. How could the movement of one minuscule particle in one of the semicircular canals of one ear be affecting all my other senses? As far as I knew, the semicircular canals had one duty—to tell us where we are in X, Y, and Z space. Yet when this tiny particle moved, it so confused my brain that it destroyed not only my sense of where I was in space. It also destroyed my ability to see, hear, and touch—along with my entire ability to sense time and space.

To me, this was the strangest thing about my condition. The entire cause of this total disconnect from reality was a problem in one sensory organ. Moreover, prior to this problem, all my senses functioned fine. So how could this one tiny particle, free floating for 7 to 9 seconds, in the fluid, in one tube, in one ear, destroy *all* of my senses? *This is the question to concentrate on.* It changed my entire understanding of science, of reality, and the five senses, including hearing.

Know the key to understanding all this is the word I've been pushing towards you throughout this chapter—the word, *conversions*. And the thing to ask yourself is how the brain converts the raw data our five senses supply into one cohesive picture of reality. And yes, this phrase is a doozy. Phrases like "one cohesive picture of reality" tend to be like this. But in this case, it's even worse because we're talking about something we take for granted every second of every day. Here, I'm referring to two things. One, all the conversion processes which feed data to our brains. And two, the brain's ability to assemble this data into a personal world view.

What I'm trying to say—and doing such a bad job at saying—is that for us, reality more resembles how fish "see" water than us sensing a literal reality. We assume what we see, hear, and touch actually exists as we see it. In truth, there is no single sense of reality. Saying this is like saying we "hear" patterns of moving air. In truth, by the time the brain gets involved, these patterns have gone though many conversions.

Do we hear vibrating skin? Do we hear vibrating bones? Do we hear the inner ear's fluid waves? Do we hear the brain's electrochemical changes? What is the reality of hearing? Does hearing—as we think of it—even exist? Or is hearing just one of our brain's watercolored illusions?

How Does the Brain Paint Our Reality?

Hopefully, it's obvious by now that as I began these experiments, my mind was filled with questions. Could it be that everything I think is real is just my brain playing Leonardo the painter? Clearly, every minute of every day, our brains paint pictures of reality far more complex than anyone has ever imagined. And I say this with such conviction—not because I'm being poetic—but rather, because I literally cannot imagine a machine which could create such an incredibly beautiful world *representation*.

Can you imagine the minds of babies, filled with swirling, raw sensory colors? All the time, their brains are trying to learn to combine all this data into a single cohesive picture of reality? To be honest, I never imagined that trying to understand deafness would make me question my very understanding of reality. This said, I now need to describe my experiments—and the facts I derived from them. I'll begin by backtracking a bit and describe my thought processes back then.

What Happens When Our Senses Don't Agree?

Okay. Data from a particle moving in my inner ear told my brain I was moving when I wasn't. My big question was, how could this one thing be affecting my other senses? To see, I asked myself which senses were in play in those moments.

One. My brain was receiving positional data from my inner ears. But the data being collected from the one funky, balance tube disagreed with the data being collected from the other five tubes. So my brain could not make sense of *where* I was in physical space.

Two. A second source of data was the sound data being gathered by the hearing portions of my two ears. This data normally gives us our auditory sense of place and space in a room. For instance, picture clapping your hands and how this sound allows you to make sense of the size of a room. But now, picture clapping while you're in a tiny closet, but hearing these claps sound like you're in a huge room.

In my case, my sense of *where* I was in space and my auditory sense of *what kind of space I was in* disagreed. So my brain could not make sense of what the space I was in was like—for instance, how big or small, whether there were walls or not, and so on.

Three. My body was trying to compensate for all this by gathering tactile data. We use tactile data to give us our sense of physical security—

the "I'm standing on firm ground" reassurance. The problem was, each time I tried to steady myself by reaching back into the bed, I could not feel anything. So while my logical mind told me this thing existed, the "reality artist" part of my brain told me there was nothing there.

Finally, my brain tried to fall back on the ultimate reality check—my visual data. What we see out there in the world creates the three-dimensional space in which the rest of the data exists. This includes our sense of distance, up and down, and left and right. But what I saw—the world was spinning, disagreed with what my mind knew—that I was lying on a bed.

The thing to keep in mind then is how my brain could not reconcile these conflicts. And it turns out, in order for the brain to create a single, cohesive picture of reality, like constellated science, it must account for all the data. In truth, this is the main rule I referred to earlier. Neither our brains—nor constellated science—are allowed to omit data in order to create a sense of reality. It just does not work.

Now consider another idea I've previously mentioned—the idea that the present scientific method requires scientists to do this. In effect, the present method makes arriving at a picture of reality more important than whether this picture is true. To do this, scientists almost always omit some degree of conflicting data, claiming it's okay. These are anomalies or placebos. Sadly, this alone accounts for why so much of what science says is true never translates to the real world.

Now consider what it was like to be me. Whenever the one tiny particle moved, suddenly, all my sources all disagreed. In effect, my senses were telling me the world out there has ceased to exist. At the same time, my eyes told me there is a world, but not the one I know. Moreover, because this made absolutely no sense, I ceased to be able to trust what I saw.

This made me feel compelled to just shut my eyes and pray for it all to go away. And this pretty much describes my first five days with this condition. I constantly tried to find ways to minimize my symptoms. I tried sleeping sitting up in a chair. I tried sleeping in a bed, but not moving. I tried not moving at all. Forget about trying to drive to work. The loose particle kept changing its resting point. And as it did, the type of head rotation which caused the vertigo to start changed as well.

In the end, I felt scared to move in any direction.

The onset of the vertigo had just become too unpredictable.

Finally—exhausted and sleep deprived—on day five, I tried one last desperate experiment. As I lay in bed on my left side, I forced myself to roll to my right. Know that as I did this, I had to draw on every ounce of courage I could muster, including that I forced myself to keep my eyes open. And as I did, I felt pleasantly surprised by what I saw. (I bet you didn't see that coming.)

My eyes told me the world was rolling diagonally—from the lower left to the upper right.

Can you picture this? If not, try this.

Try looking out at what's in front of you right now. Now imagine what you see begins to slowly roll upward—a ninety degree angle—from bottom to top. Remember, you are not moving. But what you see in front of you is rolling steadily up. Now imagine the rolling changes. It's still rolling up. But now, it rolls up from the lower left to the upper right. And when what you see reaches the upper right, this picture instantly resets back to the lower left. At which point, what you see starts rolling up and to the right, all over again.

Old Fashioned TV - Stabilization Problems

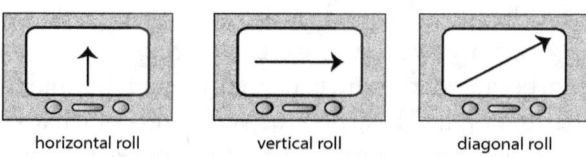

horizontal roll   vertical roll   diagonal roll

What in the world pleasantly surprised me that day?

In that instant, I recognized that pattern from my childhood. Remember, I'm really old. So when I was a child, televisions were just coming out. Moreover, those early TVs had some rather strange problems, most of which are never seen today. And one of those problems involved what the TV repairman called, the horizontal and vertical stabilization circuits.

To wit, if the horizontal stabilization circuit broke, the picture would slowly roll from the bottom of the screen to the top. And if the vertical stabilization circuit broke, the picture would slowly roll from left side of the screen to the right. Worst of all, if both broke at the same time, then both things would happen at once. The picture would roll diagonally, from the lower left to the upper right. At which point, it would reset to the bottom left—then do it all over again.

How Much of Our Reality is Real?

At this point, let me reiterate something these experiments reveal, mainly how the new method differs from the old. The new method uses an inclusive approach—the present method, an exclusive approach. For example, were I to have used the present method, I'd have limited my investigations to what my doctor had told me was wrong. Here, I'd have focused my experiments on my balance and my inner ears. But how could this have explained what was happening to my other four senses? Was something wrong with them too? For example, had I damaged my eyes? I had symptoms; the rolling picture during the vertigo.

Had I damaged my hands? I was unable to feel like I was touching anything during the vertigo. Had I damaged my ears? I went deaf during the vertigo. Had I damaged my brain? I totally lost my ability to make sense of what was going on.

In truth, to understand what was happening to me, I needed to explore all five senses. At the same time, all these symptoms actually stemmed from a single problem. A single, miniscule particle floating in the fluid in one tube of one inner ear was destroying my brain's picture of reality. As it did, even holding absolutely still would not fix this problem. Until this particle finished floating and settled, my brain could not reconcile the conflicting data my five senses were gathering.

In the end, because I knew to expand my search beyond what I could see "under the streetlight," I pushed the very limits of what I've assumed about the world of hearing, deafness, and the other four senses. To me, this is the very least a scientific method should enable you to do. Sadly, the current method not only sees nothing wrong with reducing things like the five senses to separate parts. It actually encourages—and even pressures—scientists to do this.

Not coincidentally, a recent article in *Scientific Mind* (September / October 2015) mentions this very thing. In this article, the author talks about Sensory Disintegration Disorder. Specifically, she points to how most people find it hard to believe this condition exists. Why? In part, because most people treat the five senses as if they are five separate things.

In truth, scientists cannot know the nature of deafness—or anything, really—without constellating this thing to everything around it. They must also process all of this data the same way our brains process our five senses. Our brains convert everything our five senses gather into one cohesive picture of reality. Scientists must convert all the raw data they gather into one cohesive picture of reality as well—or risk benign scientific vertigo.

Did you get this last piece?

Separating what we know about the five senses is like separating the colors in a Rembrandt—then trying to appreciate this painting. What could you know about a Rembrandt if you did this? Anything at all?

Moreover, one way to picture science is as one big societal brain. If so, then can science arrive at a true sense of reality if it separates—and even disregards—parts of what it observes?

Can it be, these reductionist separations are the fatal flaw in our current scientific method? Can this practice be a tendency built into the way our brains store things in baseline containers? Moreover, can science's failure to convert their data into one cohesive picture of reality be deafness on a grand scale? In this moment, I'm surprised at how much my sense of deafness has already changed.

As for what we should be asking about hearing as a sense, I think it's pretty obvious. Can it be that for us to hear, that all our senses must agree? And if they don't—for instance, if what we see and what we hear don't make collective sense—then what?

Is this the true meaning of the word *deafness*?

Αα	Ββ	Γγ	Δδ	Εε	Ζζ	Ηη	Θθ	Ιι	Κκ	Λλ	Μμ
άλφα	βήτα	γάμα	δέλτα	έψιλον	ζήτα	ήτα	θήτα	γιώτα	κάπα	λάμδα	μι
alfa	vita	gama	thelta	epsilon	zita	ita	thita	yiota	kapa	lamtha	mi
a	b	g, y	d	ē	z	ē	th	i	k	l	m
[a]	[v]	[γ, j̣]	[ð]	[e]	[z]	[i]	[θ]	[i]	[k, c]	[l, ʎ]	[m]

Νν	Ξξ	Οο	Ππ	Ρρ	Σσς	Ττ	Υυ	Φφ	Χχ	Ψψ	Ωω
νι	ξι	όμικρον	πι	ρο	σίγμα	ταυ	ύψιλον	φι	χι	ψι	ωμέγα
ni	xi	omikron	pi	ro	sigma	taf	ipsilon	fi	hi	psi	omega
n	ks, x	o	p	r, rh	s	t	u, y	ph	kh, ch	ps	ō
[n]	[ks]	[o]	[p]	[r]	[s]	[t]	[i]	[f]	[χ, ç]	[ps]	[o]

What Did I Learn From My New Language Experiments?

As I mentioned back in Cartesian step two, some years ago, I decided to explore how young children learn language. To do this, I decided to teach myself a language which sounded totally alien to me. Being as I only knew English and a little Spanish, there were many languages to choose from. But while I considered Russian, German, and Japanese—I eventually settled on Greek.

Strangely, I made this decision based on a single, four minute interaction. A Greek deli owner named Andy taught me to say the Greek word for "today"; simera. For some unknown reason, in that moment, I fell in love with the sound of Greek. At which point, I made my decision.

Over the next five years or so, I studied Greek on and off. At times, I actually became able to read simple sentences—on rare occasions, I thought in Greek. Most times though, I needed a mental translation step and retained very little. Moreover, as I write this, most of what I "learned" back then seems long gone.

This said, some things—such as that first word—I can still access without effort. Not only that, but when I hear that sound in my head, I hear it like a native Greek speaker would—with no translation step. I also have no trouble recognizing this sound, nor knowing its dictionary meaning. It also has a personal, situational meaning as well.

Mainly though, when I bring it to mind, I'm still pleasantly surprised. It seems, this kind of surprise may be the best proof for having learned a language—or anything, really. Indeed, in the chapter on learning, we'll explore this question in depth as well. For now, we'll stay focused on my attempts to discover how children learn language.

Know the main body of these experiments consisted of "teach yourself" Greek courses and books. Today, whenever I look at the Greek alphabet, I feel like a kindergartner—amazed by the magic of reading. I also, automatically, start reciting the Greek letter-sounds in my head. And I also still feel curious as to how many letters I can remember.

What's the Best Tool to Learn a New Language?

One aspect of these experiments involved my search for the best method. What is the best way for an adult to learn a new language anyway? During my search, I found many brands which claimed to be the best. Being most brands of Greek language software cost hundreds of dollars per level—and being most brands have multiple levels—initially, cost was a factor.

Despite this, over time, I ended up buying almost all the levels of every major brand. I literally spent thousands on software—and tried them all. Why? Because constellated science provoked childlike levels of curiosity in me. I simply could not resist.

What made me so curious? They each make different claims as to how adults learn. One brand claims adults learn best by imitating how young children learn. By this, I mean, they have you look at a series of pictures while listening to, reading, or speaking words or phrases. Notably, they only hint at meanings as they show you these pictures. Nowhere do they actually tell you what these words mean.

Another brand—one which has been around for many years—uses the old fashioned kind of rote, reductionist repetitions I experienced back in first grade. They have you break sentences apart—put them back together—then reorder them and try again. I found that when I later listened to the software, I could easily predict what was coming next. But if the word order changed to one I'd not memorized, I got lost.

Another brand uses video clips of an adult teacher teaching. She talks to you as if you are an adult in an actual classroom. This one gives you actual cultural context as you learn, along with the usual reading, writing, and speaking exercises.

Here, I felt surprised by how this helped me to absorb what was being taught. It seems, seeing this teacher's face created a more adult context. I also wonder how much seeing her mouth move added to my sense of what I heard. Evidently, no other brand sees this as important though.

In addition, each of these programs makes assumptions as to which content to focus on. Most started with various Greek greetings. Many assumed you were preparing to vacation in Greece. In essence, most Greek language software—and many Greek language books—try to teach you to survive as a tourist. No surprise, most of this stuff felt fake and forced, and none of it kept my interest.

Notably, one brand did not focus on tourist content. It focused mainly on listening to real life, adult conversations. First you listened to these conversations. Then they broke them down, word by word. Know I bought this brand only after becoming discouraged with the brands most people recommend. So much for assuming I'd learn best if I learned like a baby, or like a young child in grammar school.

To my surprise, this one gave me my second aha.

The aha? Regardless of the approach or content, inauthentic conversations failed to help me learn. Whereas content delivered in what felt like real, adult conversations kept my interest. I also loved it when they took these conversations apart, bit by bit. Somehow, this approach seemed to hook into my brain in ways my adult mind found appealing.

Realize, when I say "real conversations," I mean things like two people kidding around. Indeed, the more I felt the people had a real relationship, the more I retained. This raises what I see as the greatest flaw in language learning programs. Instead of teaching you how to have authentic sounding conversations, most language programs have you robotically parrot conversations which sound stiff and inauthentic.

Know we'll come back to this concept shortly, as it turns out it's one of the keys to understanding deafness and hearing. It also appears to hold the key to understanding the five senses in general. Here, each of the five senses create part of the context for the other four. And since all learning is state dependent—and since our minds create states of being by summing all five senses—to know a sense, we must know it in the context of the mind's big picture. This picture must include all five senses.

What Part Does Vision Play in *Hearing* Language?

A moment ago, I mentioned my surprise at how seeing a teacher's mouth moving helped me to learn. After spending a year or so studying Greek, I began to also notice Greek people's head movements as well. Here, the first time I noticed this was a time when a Greek speaker was saying the Greek word for *yes*. Bad enough this word (*nai*) sounds a lot like the English sound for no (*nay*). But when Greek speakers say *yes*, they first tip their heads down, then up—then stop.

Obviously, English speakers move their heads in exactly the opposite way. As they say *yes*, they tip their heads first up, then down—then repeat. Can you imagine how confusing this was for me? Both the sound and the head movements were backwards from what native English speakers do. Moreover, when I nodded like a Greek and said the Greek word for *yes*; *nai*, this combination created a deep conflict in me. My mind, my body, and my inner ear all told me something was wrong. But as hard as I tried, I just couldn't put these things together in real time.

Want to experience this problem for yourself? It's easy. [1] Try saying the word *yes* while nodding your head the normal way—up, then down, and so on. [2] Now try to say the word *nay* while nodding your head the same way. [3] Now try to say the word *yes* while nodding your head like a Greek; first down, then up. [4] Now try to say the word *nay* while nodding your head like a Greek saying *yes*.

Now try doing these four experiments again. Only this time, focus on what you hear. How easy is it for you to hear [1]; the normal English *yes*? How hard is it for you to hear [3]; the English *yes* while nodding your head like a Greek— first down, then up? In my experiments with people, most found it difficult to hear correctly while moving their heads backwards. Or they could hear correctly, but couldn't focus on noticing their head movements.

Similarly, if I asked them to watch me doing this, they also struggled and felt confused. Again, it seems, context affects hearing.

Now recall once more what I said a page or so back about how seeing a teacher's face helped me to absorb what I was being taught. This raises questions as to how much we need to connect language to people's facial expressions in order to hear. Indeed, what part do facial expressions play in determining what sounds mean to us? And is this why most deaf people make very expressive, facial expressions when speaking or signing? Are these facial expressions an integral part of their ability to hear?

Can You Lip-Read a Silent TV?

These questions make me wonder how much mental sound plays a role in lip reading? Can people with Type 1 deafness even lip read at all? To see, I devised yet another simple experiment you can do yourself. Turn a television on and find a talk show. Now turn off the sound and try to lip read what you see.

Most hearing people can do this fairly well. I can. But now try to lip read, but without hearing the words in your head. Can you do it? I can't. Nor it seems can most hearing people. For hearing people, lip reading seems to be connected to mental sound.

Surprisingly, studies report that most people succeed in lip reading only 30% of the time. Studies also report that lip reading plays a significant role in how all babies learn language between 4 and 8 months old. In addition, I'm told that up to the age of 14, little difference exists between the lip reading skills of hearing and deaf children. But after 14, deaf children are said to be better at lip reading than hearing children.

Then there's the "McGurk Effect," the tendency for what our eyes see to override what our ears hear. This is what makes watching dubbed movies such an odd experience. You mind can't make sense of the differences between what you hear and how people's lips move.

On the other hand, I myself love watching foreign movies with the original sound but with subtitles on. Perhaps this is because I have Asperger's and the words reach my mind more natively? Conversely, I—like most people—dislike watching dubbed movies. Indeed, the more I try to reconcile these auditory to visual contradictions, the more detached I become from the story.

This makes me wonder if dubbed movies are just another example of the failures inherent in imitating normal? Indeed, one theory of deafness—the Social Model— suggests this very thing. It claims that deaf individuals suffer more disability as a result of their environment than from their physical limitations. Experiments in this chapter seem to support this claim.

Have You Ever Experienced Tone Deafness?

Then there are people's claims about being unable to carry a tune. In extreme cases, this condition has many names—dysmusia, tune deafness, "tin ear", and dysmelodia. Whatever you call it, it seems all but a few people suffer from this to some extent.

Surprisingly, this "all" includes most professional singers.

For this reason, most recording engineers are well versed in the use of pitch correction software on voices. Used carefully, this software can improve an occasional out-of-tune note. Used more than carefully, these notes become unnatural and pitch perfect. And since no human being sings like this, these notes sound mechanically fake.

I raise this as it seems, hearing language and hearing sung melodies require different kinds of hearing. Moreover, when it comes to hearing sung melodies, it seems, many people have a degree of deafness. As a sound engineer, I can attest to this, as well as to the need to never state this directly to the person singing. At least, if you want to keep your job.

Then again, in the February 2007 issue of the Journal of the Acoustical Society of America, lead researcher Simone Dalla Bella claims 90% of people can carry a tune. She then goes on to say that those who can't can be divided into two types (there's that pattern again). Most of these folks are actually tone *deaf* and can't hear when a note is off. They *cannot hear* when they are singing poorly. But there are also people who sing badly, but *can hear* they're singing badly. Yet for some unknown reason, they cannot correct themselves. They are, in a sense, "tone *mute*."

I would bet that Dr. Bella hasn't spent time in a typical recording studio. In and of themselves, the existence of pitch-correcting plug-ins like Autotune and Melodyne attest to how often singers sing wrong notes. Moreover, these programs are expensive. Yet most small studios have at least one of these programs. From this, I'd guess the statistic should be reversed. Or at least, their definition of "being able to carry a tune" should be made more scientific. Most people cannot get through a song without singing off notes. Thus most people have some degree of tone deafness.

Have You Ever Experienced Tempo Deafness?

While most people have heard of tone deafness, most will have no clue what I mean by "tempo deafness." Yet of all the things people react well or badly to in a song, a musician's feel for the beat is one of the strongest. Yes, flat notes make people cringe and are often seen as the worst offence. But feel—a musician's sense of the timing of sound—is probably second.

This reminds me of a time I was in a recording studio. That night, the drummer and leader for the band Blood, Sweat, and Tears had asked about eight or ten of us to try to add hand claps to one of his band's tunes. Even from the first take, I could tell that some of us just couldn't keep time. Then, one by one, he asked people to leave, until it was just him and me. To be honest, I felt proud to be the last man standing. But the pressure was on and in the next take, he eliminated me as well.

So much for my audition.

Remember, too, that in those days, there were no computers to correct timing of a track. Things had to be right—or they had to be redone. Admittedly, being the last to be asked to leave did feel good. But to this day, it bothers me that I could not keep time as accurately as he could. Duh!

This is tempo deafness—the inability to extract timing information from sound. To be honest, like tone deafness, most people suffer from some of this. Considering all languages have accents and cadence, to me, this is odd. And as I think about it, this makes me wonder if this is part of what makes learning new languages hard? Moreover, does this mean we need to include accents, tempo, and cadence in our definition of hearing?

Does the Sound of Massage Oil Change the Massage?

One of my formal students owns a spa. She is also a licensed massage therapist who teaches classes for massage therapists. In my research for this chapter, I asked her a series of questions about massage. Specifically, I asked her how sound affected people's reactions to massage.

Suddenly, a new stream of questions began to emerge in my mind. She of course reported how music affects massage. But my questions went way beyond that. For instance, I asked if different viscosity massage oils (thicker or thinner) sound different when used for massage? Moreover, to what degree does this change in sound alter the effect of the massage?

My mind then went to chiropractic and to those collapsing adjustment tables some use. I can hear that sound in my head as I'm picturing getting that type of adjustment. I also prefer the type of adjustment which doesn't include that sound. Can it be the sound of the adjustment alters my experience of the good it does or does not do?

The Deafness of Imitating Normal

Finally, there's the topic I told you I'd come back to—imitating normal. I've been experimenting with this topic in my therapy practice for many years now. What does this have to do with deafness? To begin with, recall the theory I mentioned—Social Deafness theory? In effect, this theory says a deaf person's disability comes largely from the person being pressured to imitate normal. On this theory, I cannot agree more. Imitating some theoretical normal will almost always guarantee stiffness.

Then in this chapter's opening paragraphs, I referred to the things Deaf people and people like me—people with Asperger's—have in common. Do we have these things in common because we're both pressured to imitate normal. For instance, from a young age, children with Asperger's are taught they suffer from deficits, not differences. They are then taught—both directly and indirectly—that to overcome these deficits, they must learn how to do things like normal people do. Here, by "like normal people do," I'm referring to doing things in ways which do not make neurotypicals uncomfortable. God forbid people with majority personalities be tasked with having to build social bridges between themselves and people with minority personalities. After all, normals aren't the broken ones, right?

Now please allow me to apologize to you. I don't know you and you may not feel like this. But as a person with a minority personality, I've been pressured to imitate normal all my life. I've literally heard again and again that I am damaged and need to change. And in no small part, that is probably what's motivated me to construct a non-judgemental theory of human nature.

For instance, consider the effect being forced to imitate normal has had on my ability to hear. All children feel pressure to hear words correctly. People with Asperger's feel this pressure in spades. No surprise, when I was trying to learn Greek, the greatest pressure I felt was to hear and pronounce words correctly. The problem is, this pressure got so bad that it often caused my mind to go blank. In effect, I literally became deaf every time someone spoke Greek.

This pressure also prevented me from bringing to mind the meanings of words and phrases. And while some of this is to be expected, after a while, I began to feel curious as to what was happening to me. At some point, I began to listen more closely for variations in the way native Greek speakers speak Greek. I also began to pay attention to the many differences in the various language teaching programs. Here pronunciations, usage, grammar, and fonts often varied so widely that at times I had a hard time knowing what was correct.

Then, as I began to pay more attention to native speakers, I realized that none of them speak in anything like a uniform way. In other words, the *correctness* children are pressured to imitate exists only in theory. In the real world, except for the word police, no one talks like this.

To a man with Asperger's, the existence of these variations was a real revelation. When it comes to language, it seems, speaking correctly exists only in theory.

Ironically, the pressure to speak correctly may be the greatest contributor to all children's struggles to learn a language. Can you imagine the added pressure deaf children feel to hear and speak *correctly*? What about the effect being pressured to imitate normal has on learning in general? For instance, consider what happens when a child mispronounces a word or offers the wrong answer in a classroom. In most cases, at least a few children will laugh, implanting in this child more pressure to imitate normal. Over time, this pressure then generalizes to professional organizations and clubs who reinforce this need.

As for the present topic, the question which comes to mind is this. What affect does this pressure to imitate normal have on hearing peoples' ability to hear? At the very least, this pressure will divide their attention between sounds and meanings. This will then cause them to focus mainly on the first part they learned to do—hearing the sounds. This will then impair their ability to hear the meanings of these word-sounds, resulting in what I call, "semantic deafness."

Again, it appears that hearing requires both auditory and semantic hearing. And that to hear, a person's mind must combine these things into a cohesive picture of reality. For instance, right now I am asking myself, to what degree did the pressure to imitate a non existent, "normal" Greek speaker account for my struggles with learning Greek?

At this point, I don't have an answer. But I do have a lot more questions.

Cartesian Process ~ Step Four: Pattern Seeking

Finally, we arrive at the final step in constellated science's *first* process. Here we'll look for patterns in our data and then summarize these patterns. Our hope is that seeing these patterns will, in the future, give us the ability to map deafness and hearing. Please keep this goal in mind then as we progress though this step. At this point, we're not looking to make a new map, let alone parallel other maps. We're searching only for patterns within the data we've gathered.

My General Categories Listed Once More

Okay. The first thing we'll need to do is to remind you of the four general categories which have been guiding our search.

- The Physiology of Hearing: How Do We Hear?
- Sensation and Hearing: Is Sensing Sound, *Hearing*?
- The Mind and Hearing: Is Understanding Sound, *Hearing*?
- Deafness and Hearing: Do Deaf People Hear?

So what have we learned from this list? Let's see.

Cartesian Process ~ Our Working List of Opposites

Now we'll list the pairs of opposites I've mentioned in the text. We'll then explore these pair opposites, looking for parallels between the pairs. We'll then create a list of questions we can use to explore the nature of these parallels. Remember though, we're NOT looking for answers. We're looking for new lines of questioning.

Here is our working list of opposites.

- Sensory Baseline: Hearing (what is it?) vs Deafness (what is it?).
- Sensory Processing Direction: Physical-to-electrochemical conversions (sensations being converted to information) vs electrochemical-to-physical conversions (recalled information being converted to sensation).
- Physical & Mental Hearing Baseline: Physical hearing (sensing out-in-the-world sound) vs mental hearing (sensing inside-the-mind sound).
- Deafness Types 1 & 2 Baseline: Deafness / Type 1 (sudden onset, total loss) vs Deafness / Type 2 (gradual onset, partial loss).
- Semantic Deafness / Auditory Deafness Baseline: Hearing sounds without meanings (semantic deafness) vs hearing meanings without sounds (auditory deafness).
- Pitch & Tempo Deafness Baseline: Tone deafness (the inability to hear pitch) vs tempo deafness (the inability to hear time).
- Volume Baseline: Environmental Deafness (a significant inability to hear environmental sound—sounds are too soft) vs Hyperacousis (a significant intolerance to normal environmental sounds—sounds are too loud).
- Hearing-Tests Baseline: the Personal Auditory Skill vs the Personal Intellectual Skill.

- Language Type Baseline: Auditory Language (speaking / hearing) vs Visual Language (speaking / hearing).
- Theory to Real World Baseline: Hearing / speaking theoretically correct speech vs hearing / speaking actual, real world speech.
- In Season (early childhood) learning-to-hear vs Out of Season (later in life) learning-to-hear (e.g. second languages, music appreciation classes, etc.).
- Early Brain Reallocation (baseline blank space for a sense is unused / another sense uses it) vs Late Brain Reallocation (baseline space for a sense has been used / no more room for new information).
- Authenticity Baseline: The negative impact of being pressured to imitate normal on hearing vs the positive impact of being encouraged to be yourself.
- Sight & Sound Agreement Baseline: Sensory Agreement vs Sensory Disagreement.
- Vertigo Baseline: Objective vertigo (we are still & the world out there is spinning) vs subjective vertigo (the world out there is still, but we are spinning).
- Massage Focus Baseline: Physical Massage vs Auditory Massage
- Language Feedback Loops: Hearing-to-Speaking Feedback loop vs Speaking-to-Hearing feedback loop.
- Damage to Broca's area (mind okay, body struggles) vs Damage to Wernicke's area (body okay, mind struggles).
- Learning Baseline: Brain storage *containers* vs stored brain *content*.

Cartesian Process ~ Outcome: Deafness Questions

Finally, we arrive at the outcome of this particular Cartesian Process— our preliminary list of questions. We'll derive these questions from the pairs of opposites we've just listed. These questions will insure that our minds remain open during the next section's process. Hopefully, these opposites will at some point also become raw materials for new maps.

Please remember, the list of questions we're about to make takes the place of the current scientific method's hypotheses. Here, conventional science focuses—start to finish—on discovering answers. This begins with a guessed answer (the hypothesis), then explores this answer to see if it is true. Whereas constellated science focuses start to finish on discovering questions. This begins with the prototype question (what can we learn about ... ?), then expands on this question, looking to discover patterns of connections hidden within these questions.

So what questions have we discovered? Let's see.

So What Have We Learned About Deafness … ?

- To me, the biggest—and most surprising—question to emerge during this chapter is, what is *hearing*? I had no idea exploring deafness would lead to so many questions about the five senses. For example, how can we possibly know what deafness is without first having defined hearing? Yet despite repeatedly trying to arrive at this definition, as we close this chapter, we still have not done this. At least, not in a way that satisfies constellated science's standards. In part, this is due to something which emerged during the Cartesian Process—the idea that the theoretical separations between the five senses are mostly illusions. That these assumptions were mostly illusions surfaced only after I repeatedly failed to define deafness. Indeed, as I began writing this chapter, I mistakenly believed I knew what deafness is. Now, at the end of this chapter, I realize I don't know how to define any of the five single senses. Not a one.
- This leaves us with many questions, including this chapter's biggest question of all. How do we define each of the five senses?
- Can it be that the mind functions like a prism, separating the five senses similarly to the way prisms separate colors? Can each sense have a range of vibrational frequencies? Can touch be the lowest, like red light? Can sight be the highest, like violet light?
- For that matter, is there an analogous sensory state to white light? If so, then can this state be the one where all five senses blend into one cohesive picture of reality? Can it be the brain uses the five senses like primary colors as it creates this cohesive picture of reality?
- This chapter's Cartesian Process also raised many questions as to the nature of deafness? Is deafness simply an impaired ability to sense any of the five senses which involves hearing? Indeed, do any of the five senses not involve hearing?
- Then there are the functional questions regarding hearing and deafness. For instance, does the inner ear actually function like a condenser microphone?
- Is the power cycle in the inner ear an emergent property of the sodium / potassium ionization cycle? What actually interrupts this cycle? What shuts it off? What keeps it going?

- What about Type 1 deafness? Can we develop a treatment for Type 1 deafness which restarts the power to the inner ear? Could this treatment be a site-specific electro-chemical injection plus an electrical jump start? If it's not possible to restart the body's power source, could an external power source be used? Would this even matter if this deafness had gone on for too long?
- It seems Type 1 deafness has something to do with electrolyte balances. This leads to the question, does Type 1 deafness have anything in common with what causes us to get cramps? Can chemical imbalances in electrolytes in the muscles teach us more about the inner ear? Is Type 1 deafness the result of a prolonged "cramp" in the ear?
- James got Type 1 deafness at six months old, then cochlear implants in both ears at two. Brain scans show his brain lighting up as he is exposed to sound. Yet he never learned to speak. Is there any intervention which could help him later in life?
- Then there are the questions this chapter's Cartesian process raised about learning. For instance, the prototype sensory learning we call auditory hearing is certainly a type of learning. So how does the brain allocate and access what it learns about hearing?
- Similarly, the hearing I'm calling mental hearing must also be a kind of learning. Can it be, these two types of hearing (physical & mental hearing) are just different aspects of the same collective mental reality our brains construct? Can it be our minds choose to focus on one, while shutting out the rest of the collective picture?
- Then we have the containers vs content questions. To wit, can it be that early in life, the brain creates constellations of "containers" which act like keyed compartments for different kinds of life experiences? For instance, with regard to language, can one language container use the mental equivalent of shaped holes which match the elements of English, while another uses holes which match Japanese? If so, then is this what makes it so hard to retain new learning later in life? If there are no containers which fit this new learning, then it never makes it into long-term storage?
- If these containers exist, then to what degree does having them pressure us to force-fit the elements of new learning patterns into the keyed-storage compartments of already existing learning? Do these keyed storage compartments function similarly to how neurons have different receptor sites keyed to different neurotransmitters?

- To what degree can adults alter existing containers later in life? Is this the primary way we learn after early childhood? We refine existing containers?
- Do long periods of auditory deprivation—such as years of meditating in isolation or solitary confinement—reset the brain's ability to create new containers? Do people who evolve their meditation skills to a high level recreate early childhood brain states? Does this allow the brain to create new containers as well?
- Is there a doable way to create new containers later in life? For instance, can my own brain have created a new container for this scientific method? Can this have happened during the ten year period in which I managed to achieve a mentally wordless meditation practice?
- I have a high IQ, an extreme willingness to learn, an unending curiosity, and a scientific method which opens doors like they aren't even there. Indeed, this book offers proof for my ability to discover new elements of learning. Yet most of what I tried to learn in five years of highly self-motivated, Greek studies failed to stick. Why? Is my English container so fixed as to be close to unalterable in adulthood?
- If so, then why can some people easily learn new languages later in life? For instance, does culturally immersing oneself in a world wherein you do not understand the language act like a catalyst to the brain? Does doing this—experiencing so many sounds with no meanings— force the brain to create a new container to contain them? Indeed, could depriving myself of any exposure to English speakers, while also exposing myself to only native Greek speakers be the true key to my learning a new language later in life? As I imagine doing this, I am surprised to find how much I find myself loving this idea.
- Then there are the questions about the effects of imitating normal. For instance, this chapter's explorations seem to support Social Deafness theory. So to what degree does pressuring someone to imitate normal disable them? Does this then create a vicious cycle wherein the supposed solution is the source of the problem? And can this be coming from our tacit acceptance of an outdated model of disability? Can this simply be a terrible side effect of our putting too much faith in the medical model?
- One of my favorite goals as a therapist is to help people to find their calling. To me, a calling is a career which you would love

doing even if you didn't get paid. Not that you would want to not get paid. But for most people, getting paid does not make people love what they're doing. Can it be that our entire educational system is rooted in pressuring people to imitate normal? Can this be one of the main reasons so many people long for retirement?

Onward and Upward. Is This Beginning to Make Sense?

Okay. So here we are at the end of our third Cartesian Process chapter. Three down—one to go. Hopefully, by now, you're seeing the differences between the current method and constellated science. As well as some of the advantages to using the new method. And yes, I've still got a lot to explain—for instance, how to constellate maps. But at this point, I hope I've awoken in you enough curiosity to want to continue exploring further.

I'm also curious how the pressure to imitate norms may be affecting your ability to see the good here. After all, the current scientific method pressures even great scientists to comply. Grants. Jobs. Peer pressure. All conspire invisibly behind the scenes. When you then add in that our brains are programmed to accept only things which fit the holes in our brain's containers, well, it's amazing we discover anything new.

Realize the thing I'm looking to create here is what the author of the first American psychology textbook, William James, called a "cash value." To James, for a thing to be true, it must currently be making a difference. Here, *truth* to James was a living thing—and a dying thing, as well. Hence he believed something remains true only for as long as it has a cash value.

This said, I hope you're seeing the cash value in constellated science. As for this chapter's alleged theme—deafness—I hope I've stirred the pot a bit and that some of you may look more deeply into the connections we've found. And even if, like Descartes' discoveries, these connections ultimately turn out to be erroneous in the end, it's the method not the discoveries I'm lobbying for here.

Of course, I also hope that one or more of the questions I've raised here will lead to new lines of enquiry regarding deafness. There's so much more we could be doing. There's so much more we could know. There's also so much we could gain from learning to bridge these two worlds—the hearing world and the Deaf world.

Speaking of bridges, there is also the idea that it's social connections which build these bridges. This means spoken auditory language is but one bridge-building tool. Unfortunately, it seems, most hearing people take spoken language for granted—I did for much of my life. But my short time around deaf people forever changed this in me. Writing this chapter has made me appreciate the Deaf world even more.

Finally, remember, I've told you my primary job titles are personality theorist and talk therapist. In a way though, you could say that my life's work has been to create a scientific way to help people to connect. Hopefully, you and I have begun to build just such a bridge—a connection founded on a common love; the love of science. Then again, this may be redundant. After all, to me the love of science IS the love of discovering connections.

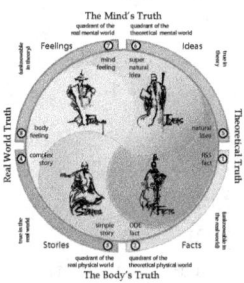

Section 2 - Chapter 13

Cancer
(what can we learn about ... ?)

Cancer and Feeling Powerless

Of all the feelings cancer provokes in people, feeling powerless may be among the worst. I felt it at age 10, when my cousin Bill—at a family picnic—told me he would likely die soon from a rare form of cancer. I felt it at age 41, when my dad—in a brief phone call—told me he had prostate cancer and had ten months to live. I felt it at age 49 when a new client—in the opening minute of his second session—told me he'd been diagnosed with throat cancer. And I felt it in my late 50's when my sister, Teresa—ten minutes into a visit—told me she had breast cancer.

Once cancer touches your life, this feeling of powerlessness never fully goes away. Indeed, I feel it even as I'm writing these words. And cancer is so powerful. What could I possibly have to say?

Fortunately, we're only looking to discover one thing—natural connections. Thus I won't be offering answers, nor advice, nor be making any claims. At the same time, most people who stray beyond the walls of peer-approved scientific research end up facing disapproval. So while I know constellated science can mitigate this a bit— still, I have to admit, I'm nervous. Then again, if I can help even one person ...

Okay. I'm ready. How about you?

Is Cancer Another of Science's Definition Problems?

Before we delve into this admittedly difficult topic, I'd first like to ask you to do a brief experiment. Imagine you're a scientist and have been tasked with exploring a new disease. To begin with, you've been given the following information. Then you've been asked if these two conditions can possibly be the same disease.

Condition one involves rapidly dividing, undifferentiated cells. These cells are located somewhere in the body of a 55 year old male. Prognosis? Left untreated, less than a year to live.

Condition two involves slowly dividing, losing-their-differentiation cells. These cells are located in the same area of the body, but in a 75 year old male. Prognosis? Left untreated, anywhere from two to ten years.

Now let me ask you the critical question. From these descriptions, can these two conditions be the same disease? Current science tells us they are—and more or less treats them as such. Indeed, most human beings treat them as if they are the same condition. Yet even a first year medical student—or a ten year old, for that matter—upon reading these two descriptions, would likely say they cannot be the same disease.

So why does science treat them as if they are?

Does Cancer Also Have a Baseline Pair?

Whenever I begin a new scientific inquiry, I like to start by looking for baseline pairs. Like finding the "corner pieces" in puzzles, baseline pairs set clear and observable limits for searches. They also insure the research will be scientifically complete—and they guarantee any facts used will be measurable in the real world. In addition, they create scientifically sound ways to define terms—and they offer broad, yet manageable starting points to begin the research.

Mainly though, they allow the mind the freedom to wonder—surely a necessity. Without wonder, science reduces to a pale imitation of itself. Rote research rarely pays off.

So does cancer have a baseline pair?

Does a bear do his business in the woods?

Many years ago, I realized, all cancers can be divided into two piles—those which grow rapidly and those which grow slowly. Indeed, for a long time now, I've been fascinated by the idea that researchers don't seem to care much about this. And yes, on occasion, people do mention this—when they're trying to determine a prognosis. But as a possible key to understanding cancer? Never. Then again, I haven't told you why this matters.

Could This Be Cancer's Baseline Pair?

A moment ago, I briefly described two diseases, then asked if they could be the same condition. One involved rapidly dividing, undifferentiated cells. The other involved slowly dividing, losing-their-differentiation cells.

I put cancers which involve rapidly dividing, undifferentiated cells into the Type 1 cancer pile. Here, it appears, something in the body has reverted to a "beginning of life" state.

Mostly younger people get this kind of cancer.

And I put cancers which involve slowly dividing, losing-their-differentiation cells into the Type 2 cancer pile. Here, it appears, something in the body has worn out and is in an "end of life" state.

Mostly older people get this kind of cancer.

The genius composer and musician, Frank Zappa, died from prostate cancer on December 4, 1993. He was 52 years young. He'd been diagnosed with this cancer in 1990. But at that time, he'd been told it had been growing for years and had become inoperable.

Was this a Type 2 cancer?

My cousin Bill got a rare form of cancer at age 26. He was born with a genetic defect and this cancer spread to his whole body in the course of a few months. He died not long afterwards and left behind a young wife and four very young children. Clearly, Bill's cancer was a Type 1 cancer.

My father got cancer of the prostate at age 73. As I said, one doctor gave him ten months to live. Yet another doctor told him not to worry—that he'd die from old age, not cancer. In the end, this doctor was right. He survived years beyond that diagnosis, this despite never having surgery. And yes, when he died at age 88, like Frank Zappa, his prostate cancer had spread throughout his body, including to his spine. But he'd lived 15 years with his cancer. Surely, my father's cancer was a Type 2 cancer.

Rapidly dividing, undifferentiated cells. Slowly dividing, losing-their-differentiation cells. That we call both these diseases "cancer" points to how poorly science defines things. Or at least, how inadequately medicine uses words. And yes, doctors treat these cancers differently. And cancer treatments get better each year. But I've seen people fall apart just from being told they have "cancer." The word is that powerful. And frightening. Moreover, studies repeatedly connect positive beliefs and attitudes to greatly improved chances to recover. So why continue to use this vague and nebulous word to refer to such different conditions? Given the devastating affect this word has on people, shouldn't we come up with a better way to talk about cancer?

Hopefully, by this chapter's end, we'll at least accomplish this.

Cartesian Process ~ Step One: Slate Clearing

As I've told you in the previous three chapters, slate clearing has one purpose—to clear our minds. We do this to bypass any and all preconceived notions as to where we should focus our research. By doing this, we stimulate curiosity in ourselves about our topic—including about where we should look. And curious minds are open minds, remember?

How do we clear our slates? We make slate-clearing lists. These lists allow us to set aside preconceived notions and clear our minds. Once cleared of assumed answers, our minds can fill with the same energy and motivation which drives all great scientists—the sincere desire to discover ways to make the world a better place.

As for making your list, all you need do is insert the topic into the prototype, slate-clearing question. In this case, our topic is *cancer*. So the slate-clearing question is: "If I had access to an all-knowing *cancer* expert, what questions would I ask?"

Now make your list.

Know it usually takes me a while to make a satisfying list. Obviously, the more effort you put in, the more discoveries you'll likely make. But in case you find it hard to begin your list, try reading mine. Then when you're done, set mine aside and make your own.

Here is my slate clearing list.

- Why does science refer to so many dissimilar cellular malfunctions (over 200) by the same word; *cancer*? What makes us call these cells *cancer*? What is cancer's essence—its sine qua non?
- Even from my brief description, it seems there may be two types of cancer—Type 1 and Type 2. If this is true, then what do these cancers have in common? And what makes them different?
- Why are cancers primarily named for where they are—their location (e.g. breast cancer, prostate cancer, liver cancer) rather than for what they are—their nature (for instance, undifferentiated, rapidly dividing cells vs slow growing, worn out or damaged cells)?
- Who gets cancer? Do women get it more than men? Do old people get it more than young people? Do light skinned people get it more than dark skinned people? And is any of this changing?
- Is there a geographic correlation to cancer? For instance, is there more risk for cancer in high northern latitudes vs near the equator? Does humidity or dryness change the risk?
- Why can't our bodies just fight off cancer? And to what degree is getting cancer just a perfect storm of bad luck? Are the cancers

people get now different from those people got fifty years ago? If so, which cancers and why?
- How many people who get cancer survive to live normal lives? Are survival rates currently going up for most cancers, or only for some? Are treatments getting better, meaning, more remissions with less side effects? If so, how much have treatments changed?
- Fetal cells are rapidly dividing, undifferentiated cells. Is there any relationship between fetal cells and what I'm calling, Type 1 cancers? Can adult cells even revert to fetal cells? If so, what kinds of things provoke these regressions?
- Exposure to high levels of radiation affects the ability of cells to divide normally. Is there any relationship between cumulative exposure to radiations and Type 2 cancers? And is this why radiation is often used to treat Type 1 cancers—to slow them down?
- Do cancers tend to run in families? Do cancers skip generations? To what degree does DNA testing accurately predict cancer? How often does DNA testing lead to false positives?
- How many cancers are linked to genetic defects? How many to environmental stressors? How many to old age? And are there genetic or environmental factors which protect against cancers?
- Can being mentally traumatized—or getting seriously injured physically—increase your chances to get cancer? Can having a stressful career up your chances? Can happiness decrease your chances? What about being married vs single? Do more single people get cancer? What about income—rich vs poor? Do people with lower incomes get cancer more often?
- What part does diet play in getting, or preventing, cancer? Can people with cancer significantly improve their chances to recover by improving their diet and exercise? If so, why does this happen?
- Is surgery the best way to keep cancer from coming back? Is combining treatments better? What are the current best treatment protocols? And are some surgeries statistically unhelpful?
- Are some cancers contagious? Can they be passed to other people? If so, how? If not, why not?
- Is there any correlation between seemingly unrelated types of preventive health care and decreased chances to get cancers? For instance, proper dental hygiene—including regular teeth cleaning—has been linked to decreased chances for heart attack and stroke. Are there similar links between self care and cancer prevention?

- How often do cancer treatments make things worse? Do all treatments have serious side effects? Are there many cases where the cure is worse than the disease? Are most treatments clearly worth the risk?
- What must happen for someone to be said to be in remission from cancer? How high are relapse rates, and do they differ between Type 1 and Type 2 cancers? For that matter, do cure rates differ between Type 1 and 2 cancers?
- Are the treatments for Type 1 cancers always different from Type 2 cancers? If so, how are they different?
- To what degree does a positive attitude affect a person's chances to recover? What are the best ways to improve one's attitude? What are the best measures for attitude?
- What is the difference between tumors which are cancerous and tumors which are benign? Can benign tumors be harmful?
- What is the role of the lymph system in preventing and spreading cancer?
- What role does the immune system play in preventing and curing cancer? Can cancer be an autoimmune disease? What about the new immune system cancer treatments? Are they safe?
- What factors do doctors look at when they assign stages to a cancer? Is this different with different cancers? Do stages always increase? Can they decrease? Can they stay the same?
- Do cancer treatments have lasting, negative effects? If so, what are they? Can they be reversed? Are they permanent?
- Can getting cancer ever be a good thing physically? Can cancer cells be used for positive purposes? If so, what purposes?

Cartesian Process ~ Step Two: Fact Gathering

We're now ready to take step 2. Here, we'll begin to explore the questions on our lists. To do this, we'll boil down our slate-clearing questions to a brief list of general categories. We'll then explore each of these categories, looking for relevant data—all the while omitting any and all conclusions, including our own.

Keep in mind, as we take this step, that our focus is on learning the method. Moreover, for the purposes of learning this method, we'll focus entirely on making categories, then collect already-summarized data. Remember, though, when doing real world step two's, you'll need to make lists of individual facts—and reference sources for these facts.

This said, here are the categories I derived from my list.

- **Definition:** What is *cancer*? Can we better define this word? More important, how can we blunt the terrible affect this word has on people?
- **Causative Factors:** Why do we get cancer? What causes it? Do we even know? How much of what we're told is pure conjecture or unscientific bullshit? Can we avoid getting cancer?
- **Effect of / on the Mind:** Studies show that mental treatments like support groups positively affect cancer outcomes. Does this mean the mind gets cancer as well as the body? Or is cancer only physical? If so, why?
- **Improving Treatments:** What lines of questioning might lead to better treatments? Are there treatments we've been overlooking? Are there treatments we've wrongly dismissed? And are there treatments we've yet to try because they exist only in theory?

Before continuing, I have to ask—how's your mind doing? Do you already feel certain you know where this chapter is going? Are you curious as to what we might find? As I've said, the thing to keep reminding yourself of is that this book is *not* about finding answers. It's entirely about finding new lines of questioning. Why keep saying this? Because answers—especially answers you're certain of—close minds. And because scientists who focus on questions experience the same motivation men like Newton and Bacon and Descartes felt—the open-minded, childlike curiosity which inspires us to make discoveries.

Foggy Room Syndrome: Symptoms and Startles

The story I'm about to tell should probably go into the Cartesian step three section; experimenting—and not into the Cartesian step two section; data gathering. In this case though, I feel it would help to mention it right up front. Admittedly, one case does not a study make. Nor do I offer it as such. But one case is still data, and one of the current method's worst flaws is its constant willingness to overlook data which doesn't fit the hypothesis.

Fortunately, we have no hypothesis to confine us here. We're looking only to collect relevant data. Moreover, as I said, this story may help us to know where to look. It involves a forty something year old woman who had been seeing me for several years. And one day, as we started a session, she asked me a strange question. She asked if the room was foggy. "Foggy," I thought? Not really. It was midday; sunlight was streaming in through a window; and there were four table lamps and two overhead lights on.

Even so, I knew better than to dismiss her question. So I began to look around the room. What was it she was seeing? I hadn't a clue. The light in the room seemed normal to me. Certainly, there was no fog. Moreover, I'd known this woman for quite a while, and I knew her to be a good reporter. In other words, I knew she was actually seeing fog. So was there something wrong with her eyesight? At this point, I didn't know.

In a later chapter, we'll explore talk therapy, and I'll elaborate on the technique I used here. I call this technique, "allergy testing," albeit, it involves no allergens. In it, you deliberately exaggerate a person's symptoms until previously unseen connections emerge. For now though, we'll limit our focus to the story and to what I did—step by step—beginning with that I asked her if we could gather some data. She agreed. But then she asked, "Do you feel hot? Could you open a window before we start?"

Can the Mind Alone Provoke Physical Symptoms?

Hot? I didn't feel hot. But as I considered her request, I vaguely recalled other sessions in which she'd asked me this same question. Feeling curious about the temperature, I left the room and checked my thermostat. It said 70 degrees. "Odd," I thought. She was seeing fog I wasn't seeing. And she felt hot when I felt fine. What the heck was going on?

I then asked her if we could try to keep going without opening the window. I wanted to exaggerate her symptoms rather than relieve them. To do this, I turned on all four ceiling lights, then asked her if she still saw fog.

Strangely, she reported, she no longer did and that the fog was gone. She also reported that the temperature in the room felt cooler—more comfortable, less hot.

Less hot? With more lights on? Truly, I had no idea what was happening to her.

I then began to experiment with various combinations of room lights turned on and off. Each time I made the room brighter, she reported feeling cooler and that the fog disappeared. And each time I turned lights off, she felt hotter and she again saw fog.

She also began to cough and repeatedly tried to clear her throat. Odd, I thought. I asked her if she needed water, but she said, no. So I continued with the allergy testing.

Am I being clear? This is all happening in a normal office room—on a perfectly bright, seventy degree, New York-in-June day—within six feet of a large window with no blinds or curtains. Moreover, it was happening to an intelligent, otherwise sane, forty-something year old woman I'd known for years. Moreover, I neither felt hot nor saw fog.

Finally, I asked her to close her eyes and to tell me how old she felt. Oddly, she replied, "two or three years old." I then asked her to try to picture where she was. To which she replied, "in a bathroom, sitting on a small wooden stool, in front of a steaming hot tub."

Long story short, at age three, she'd gotten whooping cough. To help her congestion, her mother had seated her in front of a tub filled with steaming hot water. Her mother then left her alone in that room and shut the door. And at some point, this little three year old girl got startled by being left alone in that steamy hot, foggy room.

Those who know me know my first discovery occurred in 1996. Back then, I discovered the connection between getting startled and involuntary regressions to past events. It turns out, being startled permanently associates the patterns of experience which occur in the second or two before a startle to the moment of—and aftermath of—being startled. More important, whenever people re-experience anything similar to what happened in that second or two, they regress back to—and relive—the original event.

What's also important to know is that when people regress, they temporarily lose access to any life skills they've acquired after this event. And because startles cause the mind to go blank, people have no way to know why. My client had been startled in that bathroom at age three. This had programmed her mind AND body. And each time she experienced conditions similar to what she'd experienced that day, she relived this event.

Admittedly, other than the amount of light in my room, I still have no idea what was setting her off. Whatever it was though, it seems, it was causing her to relive being in that steamy hot, fog-filled bathroom. More important, it also seems it was causing her to relive her physical state at the time. *And this reliving included the symptoms of whooping cough.*

Now as I said, one case a study does not make. Nor have I offered this story as such. I've offered it only as data in our exploration of cancer. Moreover, while at this point it may appear I've gone completely off topic—hopefully, what I'm about to say will make sense. To see what I'm getting at, please take a moment to consider what this story implies.

Altering the amount of light in my room caused a woman to repeatedly and reliably relive—and be relieved of—the symptoms of a serious childhood illness. And while there is much going on here that I cannot account for, what I do feel clear about is my method and how I used it. To wit, the only thing I did that day was to alter the conditions in my room, then ask her to observe her responses. At no time did I assume a cause, nor suggest anything to her.

Despite this, for close to an hour, I predictably and reliably elicited clear—and in hindsight, logically-consistent—responses to my testing.

Realize too that at that point, I'd spent over twenty years observing, researching, and experimenting with the relationship between startles and regressions. I'd literally witnessed thousands of examples. Equally important, I'd not found a single case wherein symptoms existed without the person's mind suddenly going blank. Nor did exploring these moments ever fail to cause a person to regress to a prior trauma.

Hopefully, by now, this story has begun to provoke some questions in you. Here, the main one should be, can the symptoms of illness sometimes be provoked by regressions, rather than by direct physical causes such as germs? The woman in the story I just told clearly did not have whooping cough. Yet she manifested the symptoms. Indeed, to this day, I am surprised by how I could get her symptoms to disappear and reappear for the better part of an hour—just by turning lights on and off.

Here then is the question I am trying to get you to consider. What is the relationship between getting startled and experiencing symptoms? More specifically, what is the relationship between reliving the conditions under which you experience a startle and a return to your previous state of mind AND body. Science has repeatedly shown that all learning is state dependent. Can startles program our cells to respond to certain conditions and in doing do, cause us to regress to a previous state of being? If so, can this previous state of being include the state of fetal development—the state wherein we all experienced rapidly dividing, undifferentiated cells?

Can startle-induced regressions be the actual mechanism beneath Type 1 cancers?

Can Some Breast Cancers Be Caused by Regressions?

Years ago, I began using thought experiments to explore these questions. At one point, I even sheepishly wrote about these experiments on my website. Back then, I spent considerable time imagining stories similar to the one I'm about to tell. My experiences with the "seeing fog" case then pushed me to go even further.

All these thought experiments followed roughly the same general line of questioning. Would the timing of a mother getting startled during pregnancy program her baby's cells to regress under certain conditions? For instance, say a pregnant woman suddenly gets cut off while driving a car. Or say she has an accident. Say too that at this point in her pregnancy, her female fetus was just about to develop breast cells—moreover, that these pre-breast cells were still rapidly dividing and relatively undifferentiated.

Certainly, we could say this startle would likely flood this pregnant woman's bloodstream with adrenaline and such—as well as altering her breathing and other physiological responses.

Now advance the calendar thirty-eight years. This time, it's the daughter who suddenly gets cut off or has a car accident. Is it possible this event could cause the same chemical cocktail to flood her bloodstream? Might she also experience similar physiological responses? And could this trigger some cells in her breasts to regress to their fetal state?

Could this kind of thing be the underlying cause of many Type 1 cancers?

Of course, this line of questioning raises many more questions, the main one being—is this kind of regression even possible? What I mean is, is it possible for adult cells to regress to the state they existed in prebirth? To explore this, I looked into the obvious source for this data—stem cell research. Not surprisingly, I found connections.

Can Adult Cells Become Fetal Cells?

Let's begin with something I hope by now is becoming obvious. All legitimate, scientific terms come in pairs of opposites. Know that stem cells are no exception. It seems, there are two kinds of stem cells as well. The first kind are called, *embryonic stem cells*. The second are called, *adult stem cells*.

Embryonic stem cells (fetal cells) are said to be *pluripotent*—they can differentiate into nearly all cells of the adult body. Adult stem cells (e.g. bone marrow, adipose tissue, blood cells) on the other hand, are said to be *somatic*. They have already differentiated. Thus they only maintain and repair the tissue in which they are found.

The thing is, scientists have managed to reprogram adult cells so that they become pluripotent. In essence, they regress to an embryonic state. Indeed, the 2012 Nobel Prize in Physiology or Medicine was awarded to Dr. John B. Gurdon and Dr. Shinya Yamanaka for their discovery that mature, differentiated cells can be reprogrammed to a pluripotent stem cell state. These cells—which are called induced pluripotent stem cells (iPSCs)—are adult cells which have been forced to express genes and factors important for maintaining the defining properties of embryonic stem cells. This process genetically reprograms these adult cells, causing them to regress to an embryonic stem cell–like state.

Admittedly, it is not known yet if iPSCs differ in clinically significant ways from embryonic stem cells. However, within this research, there exists a direct link between these "regressed" cells and cancer. Scientists currently use viruses to introduce the reprogramming factors into adult cells. And in related animal studies, the virus used to introduce the stem cell factors sometimes goes awry and causes cancers.

Another link between embryonic stem cells and cancer appears in a 2013 article written by Ian Murnaghan BSc (hons), MSc. Here scientists compared gene activity in stem cells taken from both healthy and cancerous tissue. They found that the stem cells in cancerous tissue were essentially frozen in a permanent state whereby they continued to multiply as primitive stem cells. Is this frozen state yet another indicator that these cells experienced a startle? Startles cause us to temporarily freeze.

As for whether trauma can cause adult cells to regress to stem cells, according to a 2014 article on the Genetic Engineering & Biotechnology News website, it can. Here, researchers chose to alter cells' external environments, rather than their internal workings. Specifically, they pushed the cells almost to the point of death, by exposing them to various stressful environments *including trauma*, low oxygen, and acidic conditions. And to quote the article directly, "Our findings suggest that somehow, through part of a natural repair process, mature cells turn off some of the epigenetic controls that inhibit expression of certain nuclear genes that result in differentiation" (Charles Vacanti, M.D., chairman of the Department of Anesthesiology, Perioperative and Pain Medicine and Director of the Laboratory for Tissue Engineering and Regenerative Medicine at BWH and senior author of the study).

Obviously, "including trauma" is the key phrase here. After all, my thought experiments all contained events during which pregnant women got startled. However, the idea of "low oxygen" is also interesting. Why? Because when people get startled, they temporarily stop breathing. Hence getting startled leads to temporarily-lowered, blood oxygen levels.

What Part Does the Origin Timing Play in Cancers?

Most experts agree, genetic changes cause cancer. Specifically, they point to changes which affect cell division and differentiation. In other words, cancer is said to be caused by genetic changes which affect the way cells grow and divide. Moreover, from what we've seen so far, it seems, there may be two classes of cancer which result from these changes—Type 1 cancers and Type 2 cancers.

What about the timing of these changes, though? Can we learn anything from this? For instance, science claims only a small number of cancers are caused by *inherited* genetic changes—*pre-birth* changes which originate before the person is conceived.

Indeed, most cancers are said to be caused by things which occur during the course of a person's lifetime—*post-birth* genetic changes. My first question here then focuses on the timing of these causes and on whether science's assumptions are correct. To wit, do most cancer-causing changes occur *during* a person's lifetime—*post-birth*? Or do they occur before the person is born—*pre-birth*?

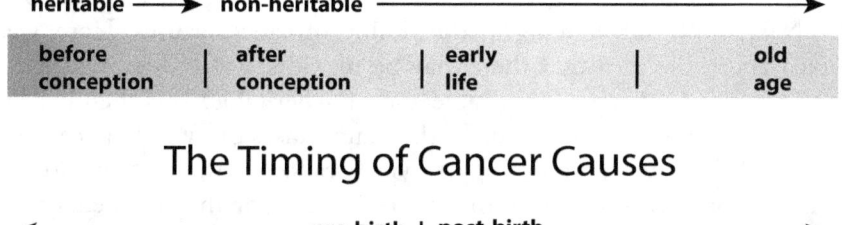

The Timing of Cancer Causes

Curiously, I've found no research which addresses these questions. Nor can I find any research which differentiates pre-birth changes which occur before conception from pre-birth changes which occur after conception. Can people be born already having experienced both heritable and non-heritable genetic changes? Can these cancer-causing flaws lay hidden until something sets them off? And can we even know what sets them off?

My cousin Bill appears to have died from the heritable variety of pre-birth changes. His genetic flaws came from something in his father's genes; something which existed before he was conceived. Conversely, the woman in my imaginary, breast cancer thought experiment would have suffered from the second kind of pre-birth, cancer-causing changes—the non-heritable kind. Her cancer would have resulted from something she experienced after conception, but before birth—a startle.

These questions reveal yet another pair of cancer-related opposites—the heritable / non-heritable, cancer-causing pair. Here both kinds of causes refer to pre-birth, cancer-causing flaws. Does this mean, both lead to Type 1 cancers? Obviously, it's too soon to ask, let alone tell.

At the same time, clearly, the timing of these cellular programming errors infers very different origins. Genetic anomalies occur before conception—startle-based anomalies occur after conception but before birth. Interestingly enough, both kinds of genetic changes may result from people having been startled. Thus another question would be, can startles lead to heritable genetic changes and if so, how often?

What Causes Type 2 Cancers?

So far, we've talked about two things—definition and timing. With regard to definition, we talked about what appears to be the sine qua non of all cancer causes—specific types of genetic changes. I'm referring here to genetic defects which affect cell division and differentiation. So does this help us to define "cancer?" Perhaps. But it seems we've a ways to go before arriving at anything like a scientific definition. Let's let it go for now.

Next, we began looking at the timing of cancer causes. Here, we talked about the idea that there may be two kinds of *pre-birth*, cancer-causing genetic changes—heritable and non-heritable. This then led to the question, are *pre-birth* changes the main causes of Type 1 cancers?

This brings us to the timing of Type 2 cancers—what's their origin timing? Moreover, is the pre-birth vs post-birth division the main delineator between Type 1 and Type 2 cancers? In other words, do the majority of pre-birth, cancer-causing genetic changes lead to Type 1 cancers? And do the majority of post-birth cancer-causing genetic changes lead to Type 2 cancers? Or is this division irrelevant—or an over simplification?

At this point, it seems, we need to look at the timing of post-birth, cancer-causing changes. Admittedly, doing this may be tough—most cancers appear only after a person is born. Besides genetics, how can we know when the problem first began? Indeed, most sources say it's rare for a fetus to develop cancer. But this, in no way, rules out post-conception, pre-birth (fetal) anomalies.

Interestingly enough, in most cases, when a fetus gets cancer, I'm told it's very treatable. But even this raises more questions. If true, then what is happening here that makes these cancers more treatable? And how can we know what is true when this question involves one of the more thorny issues in science, the one addressed by the old Latin saying which warns against assuming that—*post hoc, ergo prompter hoc*—"after this, therefore because of this."

According to this sage advice then, we should never assume that because one thing occurred before another, that the first thing caused the second. Or more to the point, as philosopher David Hume warned, we should forgo cause and effect conclusions entirely.

Where do we begin then? Perhaps, with another list.

How Many Cancer-Causing Agents Can You Name?

A good way to begin this part of the data gathering step would be to list the places we might explore. And for cancer, a good way to do this might be to name as many potential, post-birth cancer-causing agents as we can. Here again, before I show you my list, why don't you try to make your own? How many things have you been told cause cancer?

Now divide your list between the causes which are unusual, and those which are common in life.

My list? I confess. I cheated a bit. I took much of mine from the Mayo Clinic and the American Cancer Society website lists. But I then went out on my own and searched for additional causes which others claimed. This said, here's my summary of what people claim does—*or may*—cause cancer. PLEASE DO NOT SEE MY LIST AS A LIST OF WHAT DOES CAUSE CANCER. MANY OF THESE THINGS ARE THINGS WHICH MAY NOT CAUSE CANCER.

X-rays, gamma rays, exposure to nuclear weapons testing, radon—second cancers caused by cancer treatments (wow! I didn't know this)—UV (ultraviolet radiation) including excessive exposure to the sun or frequent blistering sunburns—microwaves, radio waves, other types of RF (radio frequency radiation) such as television screens, cell phones, cell towers—cigarette & cigar smoking, chewing tobacco, high alcohol consumption (because digesting it creates acetaldehyde)—living too near power lines—sitting, working, or sleeping too near certain electrical devices such as computers and alarm clocks—ELF (extremely low frequency radiation)—uranium mining, nuclear accidents, combustion engine exhausts, viruses (including 13 types of HPV), cancer-causing chemicals (carcinogens) such as asbestos and benzene—hormonal changes due to obesity, chronic inflammation, and a lack of exercise—having unsafe sex or chronic health conditions such as ulcerative colitis—eating starchy foods which have been heated to temperatures high enough to create acrylamide, eating artificial sweeteners such as aspartame and saccharin—exposure to pesticides such as DDT and lindane, defoliants such as agent orange—antiperspirants and deodorants which use aluminum, long term use of certain hair dyes, using talcum powder on the genital area, using moisturizing creams and lipstick—drinking from certain plastic bottles which have been heated, fires which burn PVCs—a specific part of air pollution known as PM 2.5 (solid dust-like particles, or 'Particulate Matter', less than 2.5 millionths of a metre across), a diet rich in pickled vegetables, such as pickled onions or piccalilli, salted fish—salt in general and smoked meats, such as pastrami or smoked beef—getting hepatitis B and C, etc.

~

Whew! Talk about feeling powerless. Just reading this list makes me cringe. It also makes me wonder how much our worrying about—or believing—that these things can cause cancer contributes to our getting cancer. In all seriousness, should I be putting "cancer thoughts" on my list of causes?

I'm also wondering if one of the truly unaddressed problems is that science looks at these "causes" individually—as single points. Earlier in the book, we spoke about how getting stuck on single points can lead to all manner of problems.

This makes me wonder if we should be assigning each of these "causes" a weighted, "may cause cancer" number. Could we then come up with an overall weighted score—an experience-based, cancer risk score?

Would this more accurately reveal the causes of cancer?

What I'm getting at here is simply this.

We spend a lot of time trying to determine whether each of these things does or does not cause cancer. But what if the things which hold the greatest danger are actually combinations and or sequences of these things? For instance, are some exposure sequences more dangerous than others, and do they ever test for things like this?

For example, do they ever test rats with combinations of deodorants, defoliants, and ELF? How about combinations of car exhaust, alcohol, second-hand cigarette smoke, and excessively loud metal music? Humor aside, at times, it seems everything can give you cancer. Yet this cannot possibly be true. Too many people do not get cancer. But it may be Type 2 cancers come from what amounts to the perfect storm of causes. I for one would love to see some data which explores this possibility.

Have You Heard of The Synergistic Effect?

Speaking of the effect of combinations, have you ever heard of the *synergistic effect*? I encountered it years ago when I was studying for my state's Alcoholism & Addictions credential. Indeed, one of the more surprising things I learned back then was that some combinations of drugs and alcohol cause the side effects to multiply, rather than add.

For example, say a person has on many occasions consumed an excessive amount of alcohol and been okay. Say also that this same person has on many occasions taken a Xanax or a Valium and been okay. The thing is, taken together, the effects of these things may not add together—they may multiply. Thus the effects of drinking alcohol and taking a Valium may be many times what the individual effects summed together would be.

My question here is this. Can the synergistic effect apply to cancer causes? In other words, can things which—by themselves—do not cause cancer, cause cancer when combined with other things? Moreover, what would we need to do in order to test for this? Would we have to test for how safe or dangerous each and every combination is? Is doing this even possible?

Finally, what about the difference between acute and chronic exposures? Most cancer testing only assesses acute cancer-causing agents. What about long term exposures? Can we be missing things here?

Can Cancer Be An Emergent Property?

Back in Book I, we spoke about the idea of *emergent properties*—an incredibly complex and intriguing concept to be sure. In essence, emergent properties are those properties of a complex system which appear only after the parts are joined in just the right way. Not coincidentally, I mentioned one of my favorite examples a moment ago—the idea of "the perfect storm."

Know that this implies the reverse is true as well. What I mean is, if you divide the joined parts of a complex system back into separate parts, the emergent properties of this joined thing will cease to exist as well. If so, then like learning, emergent properties are state dependent. By this, I mean these properties exist only in certain "joined" states.

In science, there are many examples of joined things having emergent properties. One of my favorites involves ordinary water. Here, oxygen (O) and hydrogen (H) are both gasses. But water (H_2O) has the property of being "wet." Moreover, the property of *wetness* emerges only when the H and O are joined in just the right way. At the same time, separating the H and O causes this wetness to disappear.

The question, of course, is, where does the quality of wetness come from? Certainly, wetness does not result from anything logic could predict. In part, this is why constellated science so often warns against using logic to understand nature. Indeed, this caution against using logic to know a thing's nature is so important, it's given special attention. For instance, it's one thing to dissect a living frog into organs (ugh!)—quite another to think you can reassemble these parts back into a living frog.

For the frog then, being alive is an emergent property. In other words, this condition exists only when the parts are joined in just the right way. Indeed, it seems, this same idea applies to all organic life forms. Life itself is an emergent property.

Scaled down, this same idea also applies to the cells in our bodies. They are alive only when the component parts are joined in just the right way. Cancer modifies some of the main components of cells—it modifies certain genes. This means, while the component parts of cancer cells are still joined and alive, the properties which make them alive differ markedly from those of normal cells.

For instance, instead of reproducing only when needed, respecting neighboring boundaries, and dying when damaged or too old (apoptosis)—cancer cells reproduce when not needed, do not respect neighboring boundaries, and can override signals that tell them to self destruct when they should. Cancer cells also lose a property called, "stickiness," the idea that the surface of normal cells contains certain molecules which make them physically "sticky." This property causes cells to stick to adjacent cells, keeping them in just the right place.

The surface of most cancer cells lack this quality.

Yet another emergent property of normal cells is that they send and receive chemical signals. These signals tell cells when they've reached their growth limit and to stop growing before causing damage. The thing is, while normal cells obey these signals, cancer cells ignore, misinterpret, or exaggerate these messages.

Of course, if we extend the idea of emergent properties out to include the sum of all of our cells—cancerous or not—this sum is what we call, our bodies. Thus the idea that we have bodies is also an emergent property. In effect, our bodies exist only when the component parts are joined in just the right way. Alter these parts and our bodies begin to malfunction and disintegrate.

My main question here is, can cancer be a regression to a time before an adult state of health emerged? And does cancer lead to the loss of certain emergent properties necessary for health, while at the same time causing others to emerge?

More on this in a bit.

Are Tumors Failed Attempts to Form Organs?

Yet another pair of opposites exists in the way cancer spreads. This time I'm referring to the way some cancers form tumors—and some metastasize. What I'm saying is, healthy adult cells don't just locate themselves anywhere they want to go. They each have places they belong in, and the stickiness I mentioned a moment ago is part of this.

Even tumors function somewhat like this—tumors are self-contained collections of cells. Here, each of these cells "belongs" in a primary site—in this case, within a tumor, within an organ. In some ways then, tumors act like malevolent organs within organs. The cells in a tumor stay within the boundaries of this tumor. And the tumor stays within the boundaries of an organ.

The opposite is true for cancer cells which metastasize. These cells seem to completely disregard normal cellular boundaries. This leads me to wonder if tumors are a cell's failed attempt to form another organ?

With regard to metastasized cancers, can metastasized cancer cells be Type 1 cancer cells which suffer additional, genetic signaling damage? Here the signals I'm talking about are the ones which tell a cell where it should be. So can metastasized cancer cells be cells which have regressed to the time in fetal development wherein there were no organs?

Moreover, if metastasized cancer cells do suffer more signaling damage, does this additional signaling damage make individual metastatic cancer cells harder to kill than the individual cells contained in primary-site tumors? Indeed, what exactly makes self-contained cancer cells different from metastatic cancer cells? More on this a bit later.

Take cancer cells in a primary site. Can these cells still be the kind of cells in this organ (e.g. lung cells, heart cells, liver cells)? But instead of mature cells, can they more resemble cells at the time when this organ was just beginning to form? And what about cells which have begun to metastasize. Can they be cells which have regressed even further? In other words, can metastasized cells have lost so much of their differentiation that they no longer resemble cells from any specific organ?

Ionizing vs non Ionizing Radiation

One of the more obvious classes of potential cancer-causing agents is radiation. Radiation is waves of energy radiating out from a source. Not surprisingly, when it comes to radiation-based, cancer-causing agents, we find yet another pair of opposites. In this case, experts divide potentially cancer-causing radiations into *ionizing* and *non ionizing* radiations.

Ionizing radiations have enough energy *by themselves* to break chemical bonds, directly damaging the DNA inside cells and causing cancer.

Non ionizing radiations do not have this power.

X-rays, gamma rays, and ultraviolet (UV) light are examples of ionizing radiations.

AM & FM radio waves, microwaves, visible light, heat, and other RF (radio frequency waves) are examples of non ionizing radiations.

Of course, even non ionizing radiations can cause cancer if you get exposed to large enough doses. So has science made a grave error claiming these things are either cancer-causing or not based on normal exposures to individual agents? In other words, should we be including what I mentioned earlier—cancers caused by the perfect storm of less dangerous radiations? And if so, how would we even begin to measure this?

What About Cell Phone SAR (specific absorption rate)?

Speaking of radiation, most of us know that cell phones emit a kind of radiation—RF energy. Moreover, the amount of RF energy a person's body absorbs from a cell phone even has name. It's called the specific absorption rate (SAR). In addition, different cell phones have different SAR levels, and cell phone makers are required to report the maximum SAR level of their product to the US Federal Communications Commission (FCC). They also set an upper limit. The upper limit of SAR allowed in the United States is 1.6 watts per kilogram (W/kg) of body weight

The thing is, according to the FCC, comparing SAR values between phones can be misleading. The listed SAR value is not usually based on what users would typically be exposed to with normal phone use. It's based on the phone operating at its highest power. Indeed, a number of factors affect the actual SAR value during use. So it's possible a phone with a lower listed SAR value might actually expose a person to more RF energy than one with a higher listed SAR value.

What does any of this have to do with cancer? Let's see.

So Do Cell Phones Cause Cancer?

As noted above, the RF waves given off by cell phones are non ionizing—alone, they don't have enough energy to damage DNA directly. Because of this, many scientists believe cell phones don't cause cancer. And most studies done in the lab support this theory. For example, several studies in rats and mice have looked at whether cell phone RF might increase the chances that other known carcinogens (cancer-causing agents) will lead to tumors. These studies did not find evidence of tumor promotion.

On the other hand, scientists report that the cell phone RF affects human cells (in lab dishes) in ways which may help tumors grow. Indeed, the US National Toxicology Program is currently doing a large study to see if exposure to RF energy could lead to health issues. These researchers will expose large groups of lab mice and rats to RF energy for several hours a day, for up to 2 years. They'll then follow (observe) the animals from birth through old age.

In the meantime, a recent small study in people has shown that cell phone RF may have some effects on the brain. The thing is, it's not clear if these effects are harmful. For example, this study found that when people had an active cell phone held up to their ear for 50 minutes, brain tissues on the same side of the head as the phone used more glucose than did tissues on the other side of the brain.

What does this mean?

Glucose is a sugar that normally serves as the brain's fuel. Moreover, glucose use goes up in parts of the brain when this part is in use, such as when we are thinking, speaking, or moving. So cell phone radiation does affect the brain. But the possible health effect, if any, from the increase in glucose use from cell phone energy is unknown. And unknown is a scientific code word for "we need more data."

Cancer Cell Migration vs Multiplication

My last example of data gathering involves a rather interesting article. This article was published in the journal *Developmental Cell* in October 2015. It describes a study which, for the first time, uncovers the genetic mechanism which determines whether cells will migrate or multiply. More important, it goes on to say, these two processes are mutually exclusive. In other words, at any given time, cancer cells can either multiply or metastasize—*but not both*.

As for the difference between these two things, most people know that metastasizing cancers are generally seen as more dangerous than localized cancers. Metastasizing cells don't sit still in one spot. Instead, they *metastasize*, meaning they migrate from their original sites. They then establish new tumors in other parts of the body. And once this happens, obviously, the cancer is far harder to eliminate.

This idea seems to be what was driving this study. They wanted to know what makes metastasizing cells different from non metastasizing cells. So what did this study find? That to invade neighboring sites, cancer cells must stop dividing. In other words, these two processes—invasion (metastasizing) and proliferation (rapidly dividing)—are mutually exclusive processes.

This makes these two things yet another pair of nature's complementary opposites. In effect, these two radically different states of cancer cells amount to two completely different diseases. This makes me wonder, can it be, these two processes act like waves, alternating between proliferation and invasion? Or can metastasizing cancers simply be a regression to an earlier version of fetal cells, perhaps to a point before the genetic development crosses a certain tipping point?

Then again, can these two states just mirror the two states fetal cells can occupy? Move to a new spot—create a new organ? Move to a new spot again?

Here we go again. Questions, and more questions.

Cartesian Process ~ Step Three: Experimenting

In step three—experimenting—we'll be looking to discover new facts. Specifically, we'll be looking to discover things which make us even more curious. Remember, to be a fact, a thing must reference three qualities: a specific time, a specific place, and a specific observable event. Omit any of these three qualities and a thing is not a fact.

At the same time, to create an experiment, we must also be able to describe this experiment with ideas. Ideas are summaries of correlated experiments, and experiments are meaningful sequences of facts. Keep in mind then that to succeed, an experiment must result in facts. So while your ideas may be well grounded in science, even well grounded ideas are not facts. And all data must begin with facts.

A Few Words on Designing Experiments

Something I have yet to address involves experimental design. For example, obviously, I'm not a medical researcher. So with cancer, there are limits to what I can explore. Know that addressing these limits is important, as it raises yet another question about doing science. In cases like this, how do you satisfy this step of the Cartesian Process?

In constellated science, all experiments must include both physical and mental data. Both types of experiments are valid. Both kinds of data are useful. Ultimately, all that matters is that an experiment leads us to new lines of questioning. If it does, then constellated science sees this experiment as legitimate—and successful.

My point? At times, the outcomes of thought experiments can be just as valuable as literal, physical experiments—sometimes, more. But even thought experiments must eventually reference actual, physical data. For example, take my cancer thought experiments with pregnant women in car accidents. These thought experiments consisted entirely of mentally exploring already known—but previously unconstellated—data.

At the time, this allowed me to bypass the need to acquire additional, physical data. But when the outcomes of these experiments led to questions, I then had to find actual physical evidence. For instance, as I mentioned, my thought experiments centered in part on whether trauma could cause adult cells to revert back to stem cells. And when I looked for physical data to support this, the results were clear.

Already existing experiments say trauma can cause this to happen.

As for the mental aspect, I referenced my more than two decades of work in and around trauma—specifically how trauma can program the mind and body to regress. The constellated thought experiment then explored a single idea—that if an adult relives a fetal trauma, that this might regress certain adult cells to stem cells.

The point o f course is that while I was unable to directly experiment on the physiology of cancer, this inability did not prevent me from doing cancer-related experiments. At the same time, this raises yet another question. How many experiments must you include to satisfy this step of the Cartesian Process?

My response? To be honest, I'm not sure I know. But for this chapter, besides the two I've already mentioned, I did two more.

Both experiments involved women with breast cancer.

Both experiments focused on their mind's effect on their cancer.

Admittedly, both of these experiments would fail to meet the conditions conventional scientists require. Fortunately, they more than meet the conditions constellated scientists need. The conditions? Both experiments observe the nature of cancer in the real world, including a comprehensive set of real world variables. They also include at least one tipping-point based measure to insure measurably significant outcomes.

What did I find? Here's the first experiment.

My Sister and the Chemotherapy Experiment

In the opening paragraph of this chapter, I mentioned my sister getting cancer. Know that when I visited her that year, at some point, she went for chemotherapy. My sister—being a character type two—told me I did not need to go. But being a two as well, I told her, ha. Fat chance.

That day, when they called her into the chemotherapy room, she again told me I didn't need to come in with her. But of course, as she got up, I followed her in. Then while they hooked her up, I sat next to her. And even today, I can picture a room filled with bright sunlight and chairs like blood donors lay in. And in one chair was a woman a bit younger than my sister.

At some point, I remember thinking about studies of cancer support groups. These studies all agreed, attending these groups upped one's chances for a successful recovery. With this in mind, I soon found myself informally leading a cancer support group—without calling it this, of course. And I began with the woman who'd preceded us in.

Within minutes, we had one heck of a discussion going—quite intimate, in fact. And while my sister didn't say much, I could see by her eyes, she was alert, connected, and listening. Then another woman then came in and I soon got her to join in as well. Then another came in and the first woman left, and we all continued talking.

What did we talk about? Everything cancer related. The shock of being diagnosed—the fear of death—how the suffering was affecting their families. Realize these talks were occurring in Charleston, SC, a culture which abhors personal revelations. Even now, I am surprised by how willing these women were to talk about their cancer. Grateful, even.

Admittedly, during this discussion, I momentarily felt less powerless. It seems words openly shared have the power to mitigate even these feelings—even for a family member. My overarching question at the time though was far more ambitious. Could this discussion affect my sister?

Know the outcome surprised even her oncologist.

What happened?

About two weeks later, my sister visited her doctor for a checkup. This visit included the usual exam and getting the results of blood tests. Surprisingly, during this visit, her doctor said her recovery had suddenly sped up. Moreover, nothing he had done could explain this.

Of course, when my sister came home, she was happy and relieved and such. But when she called and told me, I couldn't help wondering if our little informal group had anything to do with this. That it happened is a fact. Her doctor's comments more than point to this. But what could account for this sudden, notable change in her recovery? I cannot say.

Why mention this event then? For one thing, because it's a good example of the kind of experiment one can do—given little to no medical access. More important, it meets constellated science's requirements for a valid experiment. In it, I observed the nature of cancer in the real world, including a comprehensive set of real world variables. I also referenced a tipping-point based measure—was the outcome notably different than expected? And in this case, as I said, it was.

Admittedly, to this day, I have no way to prove a connection. And again, one event a study does not make. Still, it's data, plain and simple. And a genuine scientific method must at least try to account for it. More important, this data cannot simply be disregarded.

Dee Avoiding the Med Stop

My second experiment involved a middle-aged woman who survived breast cancer. When she and I met, she had been cancer free for almost three years. Despite this, she spent parts of every day afraid her cancer would return—not uncommon. In fact, early on, she confessed she couldn't even drive on the street where she'd been diagnosed without reliving the pain of being told. Indeed, since that day, she had done her best to avoid driving on that street.

All this points, of course, to that she had been startled by being told she had cancer. And like all startles, this event had frozen her mind in a painful moment. After that, there was nothing in her life this fear did not affect. Her fearful thoughts had become a cancer in her mental life.

What's interesting here is how this mental injury resembled her cancer. If true, Type 1 cancer cells freeze in a fetal state—and her mind was frozen in a painful moment. Long story short, within weeks of her becoming able to comfortably drive on that street, her whole life changed for the better. Indeed, her fear lifted so well, she began volunteering at a local cancer support group. This from a woman who couldn't say the word *cancer*.

Now, over a decade later, she is still cancer free. Can this have helped?

Cartesian Process / Step Four: Pattern Seeking

Finally, we arrive at the final step in constellated science's Cartesian Process. Here we'll look for patterns in our data. Our hope is that these patterns will lead to a new map. Please keep this in mind then as we progress though this step. In this step, we're not looking to make a new map let alone parallel other maps. We're searching only for patterns within the data we've gathered.

My General Categories Listed Once More

Okay. Here are the four lines of questioning we started out with—the four groups of questions we generalized from our slate clearing list. Again, we were only looking to expand on these questions, without seeking answers. So have we? Let's first look at the list.

- **Definition:** What is *cancer*? Can we better define this word? More important, how can we blunt the terrible affect this word has on people?
- **Causative Factors:** Why do we get cancer? What causes it? Do we even know? How much of what we're told is pure conjecture or unscientific bullshit? Can we avoid getting cancer?
- **Effect of/on the Mind:** Studies show that mental treatments like support groups positively affect cancer outcomes. Does this mean the mind gets cancer, as well as the body. Or is cancer only physical? If so, why?
- **Improving Treatments:** What lines of questioning might lead to better treatments? Are there treatments we've been overlooking? Are there treatments we've wrongly dismissed? And are there treatments we've yet to try because they exist only in theory?

Cartesian Process ~ Our Working List of Opposites

Now let's briefly list some of the pairs of opposites we've discovered during the chapter. We'll then create a list of questions which can be used to further explore the nature of these parallels. Again, we are NOT looking for answers. We're looking only for new lines of questioning.

Here are the pairs of opposites I've found so far.

- Type 1 cancers (rapidly dividing, undifferentiated cells) vs Type 2 cancers (slow growing, worn out or damaged cells).
- Cancers which cause cellular regressions vs cancers which cause cellular damage.
- Static (primary site) cancers vs migrating (metastasizing) cancers.
- Individual cancer-causing agents vs constellations of cancer-causing agents.
- Mental, cancer-causing agents vs physical, cancer-causing agents.
- Genetic (heritable) vulnerabilities to cancers vs environmental non heritable) vulnerabilities to cancers.
- Embryonic stem cells vs adult stem cells.
- Fetal cells (progressive development) vs Type 1 cancer cells (frozen development).
- Ionizing radiations vs non-ionizing radiations.
- Single-agent (unconstellated), ionizing radiations vs multiple-agent (constellated), ionizing radiations.
- Mind-oriented cancer treatments vs body-oriented cancer treatments.
- Single-focused cancer treatments (e.g. chemotherapy, attending a support group) vs constellated cancer treatments (combinations).
- Pre-birth / pre-conception (inherited), cancer-causing genetic changes vs post-birth / post-conception (trauma related), cancer-causing genetic changes.
- Pre-birth, cancer-causing genetic changes vs post-birth, cancer-causing genetic changes.
- Cancer treatments which enhance the body's natural ability to fight cancer (e.g. immune system enhancers) vs cancer treatments which augment the body's natural ability to fight cancer (radiations and chemotherapies).
- Cancer treatments which affect rapidly diving cells vs cancer treatments which damage and kill all cells in a specific area.
- Cancer treatments which employ chemicals vs cancer treatments which employ radiation.

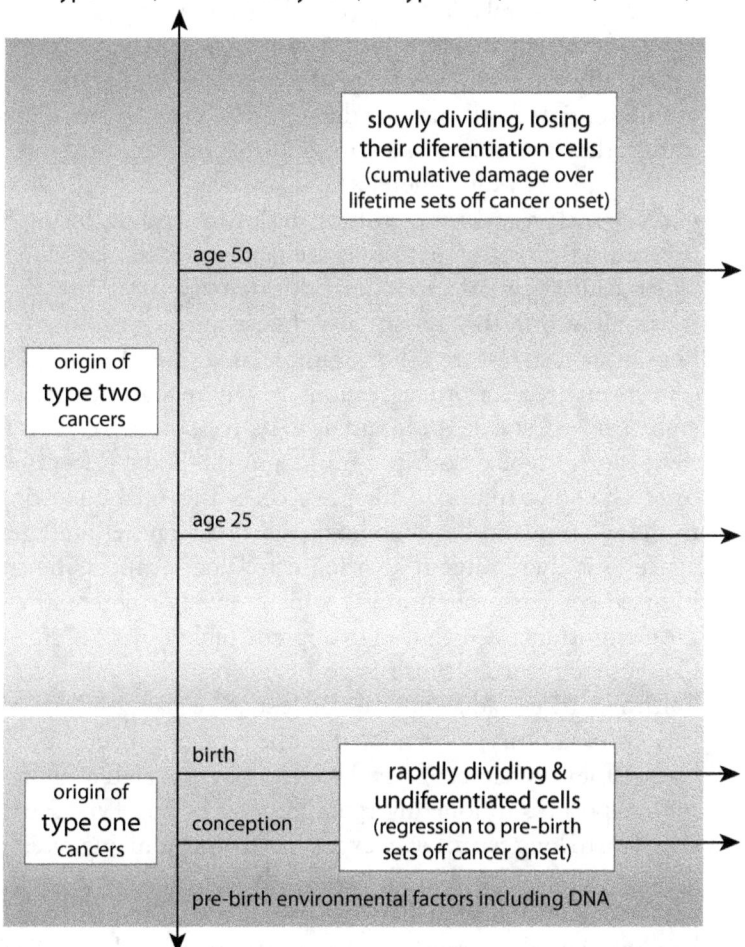

So What Have We Learned About Cancer … ?

Finally, here are a few of the questions which have emerged during this fourth example of constellated science's first process—the Cartesian Process. As I've said, we're not interested in arriving at answers, only in discovering new lines of questioning. Admittedly, doing this with such a complicated topic is difficult at best. Moreover, we've barely broached the subject. This said, we have indeed arrived at a few new lines of questioning.

Can any of these new lines of questioning lead to productive new areas of cancer research? I, for one, certainly hope so.

The lines of questioning?

- Clearly, the central question is, can there be two types of cancers—Type 1 and Type 2? Here Type 1 cancers would be cellular-level, genetically induced, *pathological regressions* to a state wherein undifferentiated cells rapidly divide. These cells are not damaged. Rather they are frozen in a temporally immature, heritable state. Type 2 cancers would then be cellular-level, genetically induced, *pathological breakdowns* resulting in slowly dividing, losing-their-differentiation cells. These cells are genetically damaged in ways which cause new cells to inherit this damage.
- If we allow for this possibility, then this suggests many new lines of cancer research. For example, take the idea that Type 1 cancers may result from regressions. My more than two decades of studies reveal, startles program us to regress; moreover, that these regressions involve both the mind and the body. Recent cancer research shows trauma can regress cells. This raises questions as to the relationship between being startled (trauma) and getting cancer—is there one? If so, then can Type 1 cancers be startle-induced cellular programming which—when triggered by prior life circumstances—can lead to unpredictable genetic regressions? Can reliving startles cause Type 1 cancers?
- With regard to radiation, can the cancer-causing effects of the various radiations be correlated to the type of cancer they might cause? For instance, can Type 1 cancers be correlated to short term, high exposures to ionizing radiations—whereas Type 2 cancers correlate to long term, lower exposures to non-ionizing radiations? Or can cancer-causing radiations only lead to Type 2 cancers?
- Can radiations have a baseline pair? If so, then can we treat cancers caused by one type of radiation with this radiation's complementary opposite?
- Immunotherapies seem to hold great promise for treating some cancers. Does this imply that treatments for autoimmune diseases, such as RA, may hold similar promise for treating cancer?
- What would science find if it explored the synergistic effects of various combinations of potential cancer-causing agents? Are there dangerous combinations? And do some things which science has ruled out as cancer-causing agents actually contribute to the tendency of other cancer-causing agents to cause cancer?

- If science were to constellate potential cancer-causing agents while looking for synergistic effects, could we come up with a realistic set of real world measures which people could use to reliably lower their chances of getting cancer? For instance, would one scale tip between acute and chronic agents? Would another tip between single agents and constellations of single agents?
- Could we also come up with a realistic set of real world measures which people could use to improve their chances to recover from cancer? For instance, are there constellations of cancer-inhibiting agents which could help without also causing harm? Are there climate-related environments, such as areas with high sunlight or low temperatures, which could improve cancer treatments?
- It seems that cancers which are limited to a primary site tumor and metastasizing cancers are yet another pair of complementary opposites. Recent studies reveal that cancers can either be pathologically replicating OR pathologically migrating—but not both. What makes tumors self contained? Is it simply the as yet unknown properties which prevent cancer cells from migrating? And what makes cancer cells migrate beyond the primary site? Moreover, to what degree is this same pair of opposites at work in fetal development?
- Obviously, the word *cancer* leads to detrimental effects on people? Knowing this, wouldn't it be better to find a more precise naming convention? If so, how would we approach this? And would this require a complete revisiting of medical diagnostic science?
- Clearly, studies show positive attitudes correlate to successful cancer treatments. Likewise, sharing struggles with others (e.g. in cancer support groups) positively correlates to successful cancer treatments as well. But if we set aside outrageous claims as to what the mind can accomplish, what are we left with? More important, what exactly is the mechanism by which the mind is causing these affects?
- Lastly, I want to mention something which has eaten at me throughout this chapter. Again and again, I've questioned the connection between regressions and cancer. In my sleep, I've found myself wondering if we should be looking for the mechanism which progresses stems cells into adult cells. More important, if we found it, could we use it to progress Type 1 cancer cells into healthy adult cells? And would this be the ultimate cancer treatment—the cure which would finally cure cancer?

Closing Thoughts on the Cartesian Process

When I began writing this section, I had no idea where it would lead. My editor and I chose four topics to explore, each based on small discoveries I'd made over the years. In all honesty though, I was nervous. What if the new method failed to work? Now here I am—almost a year later, relieved and surprised by what I've learned.

My favorite discovery? That my claims about the new method are real. It's worked even for topics which I have had no formal training nor prior knowledge of. In truth, it works so well that at this point, I can't imagine using any other method. Moreover, after watching my editor inadvertently make her own discoveries just from editing this text, it appears to be a method others can benefit from as well.

Can you believe, even as I'm writing this closing, things are still coming to me? A moment ago, I wondered as to whether the new method might apply to education as well. Can it? To wit, if I am right about the nature of learning—that in order to become more than mere parrots of other people's learning, that we each must reinvent the wheel—then constellated science may be the method educators have been seeking for millennia. Could this be true?

What about our focus in this book—making scientific discoveries? Can it really be as simple as asking questions which lead to new lines of questioning? And can we really reclaim the open mindedness we once had as young children? Conversely, can focusing on answers be the main flaw which has limited our science? And yes—obviously—answers can make great stepping stones to new questions. But as an end in and of themselves? Or can it be, answers exist only in relation to where our minds are currently standing—and that they move every time we move? In effect, can answers be the mental equivalent of the ends of rainbows?

As for the lines of questioning we've explored in this book, clearly, much work remains. Can any of these new lines of questioning lead to better treatments or prevention? I so hope so. And what about you? Do you feel more curious about the world than when you began? If so, then please do let yourself continue to explore. But if not, then perhaps you could benefit from scientifically exploring your own certainty. After all, one can never acquire too much self knowledge. Moreover, if I'm correct about the nature of the certainty / curiosity pair, the future of science itself rests largely on how well we can manage to see past the current scientific method's closed-mindedness.

Is this even possible? Can we learn to reopen our minds?

To me, this may be the greatest question of all.

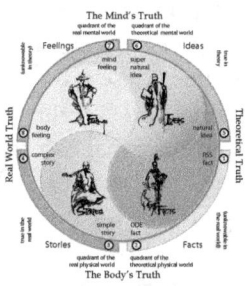

Section Two - End Notes (10 thru 13)

Afterthoughts & Resources

Notes Written in the Margins of Chapters 10 thru 13
Chapter 10 - Sleep

On The Tao and Seeking Questions, not Answers

At some point during the writing of chapter 10 (Sleep), I took a break. I then picked up a small book on philosophy and randomly opened it to a page. On this page, the author talks about Lao Tzu and his book, the Tao Te Ching. Specifically, he talks about how the Tao urges us to avoid using reason to understand nature—that thinking is the root of all problems.

He then goes on to say that the Tao urges us to not pursue "goals." Rather, we should immerse ourselves in nature itself—that the path to immersion is non-striving; simply living in tune with nature. As I sat considering this, I began to see the Tao in a new light. I've studied the Tao for much of my life. But I never realized it refers to Lao Tzu's scientific method. And yes, what Lao Tzu advocates for is more the science of creating a good life than science for science's sake. Still, the entire focus of the Tao is exploring nature's pairs of complementary opposites as the means to a better life.

As such, constellated science owes a great debt to Lao Tzu.

Chapter 11 - Weight

On Bulimia, Diagnosis, and Focusing on Personal Cause

When I planned the chapter on weight, I knew I'd need to include anorexia and bulimia. Together, these two conditions describe the two extremes in and around eating and weight. Sadly, in the course of my research, I repeatedly read professional assertions inferring people who are anorexic and bulimic do this to themselves. Why? These professionals claim these folks fear getting fat—or see themselves as already overweight.

Then at some point, I flashed back to a time when a woman with bulimia called me. We'd never met—and she'd been struggling with bulimia for twenty years. Moreover, because she lived almost two hours away, I began to ask about her life. I wanted to be sure if she traveled to see me, that I'd have something to offer.

Oddly, it took only minutes to discover the nature of her problem. All I'd needed to do was to ask her a single question. I asked her if she could ever remember a time when she was not bulimic. Somehow, in twenty years of treatments, no one had thought to ask her this question.

Does this sound odd to you too? This woman had been hospitalized many times. After each hospitalization, she'd get out and be better for a while. And of course, she'd been told—and had come to believe—that she feared getting fat. But twenty years of telling her this had not helped.

I then asked her to tell me about her first bulimic episode. No coincidence, it had occurred exactly twenty years before. It had come on unexpectedly during a Thanksgiving meal which had started out fine. But at some point, she'd suddenly felt sick, ran to the bathroom, and threw up.

After that, this pattern had repeated itself, every year, for twenty years. Most years, she had to be hospitalized, always around the same time of year. Moreover, according to conventional science, this was being caused by a fear of getting fat. Yet not once had this woman mentioned her weight to me or that she was fat.

Know I followed her lead on this and didn't ask about her weight. Rather I focused my questions entirely on gathering data about her life. The data I found most interesting? What had happened to her right before—during—and right after that first incident. Not surprisingly, within minutes, I saw what twenty hospital stays had missed.

Can you guess yet what had been provoking this woman's bulimia? How about if I give you twenty guesses—do you think you could get it then? How about ten thousand guesses? Do you think this would do it? Here's a hint. Were you to focus your efforts on finding a logical cause for her condition, you would likely spend the rest of your life making efforts and still never help her. And yes, she had come to fear getting fat and all the other psychobabble nonsense. They'd burned this into her. Unfortunately, these beliefs had only buried the real nature of her problem further.

How did I discover it? I followed Lao Tzu's advice. I avoided reason and instead, focused only on gathering—then constellating—data. And yes, she and I spoke that day for almost two hours. And she came to see me once after that. But the real work took less than twenty minutes.

Are you getting curious? Would you be mad at me if I didn't tell you? Okay. I'll give you another hint. My favorite maxim—*you can't change what you can't see*. So what had no one pictured? The real event which connected her relapses to holidays. Sadly, all her previous therapists assumed it was Thanksgiving *overeating* that was triggering her.

In truth it had nothing at all to do with a fear of getting fat.

Finally, I need to mention one more thing before I tell you. This thing involves yet another of modern psychology's wrong-headed "assumptions." The assumption? That people overeat to stuff their feelings. The truth? People who overeat—for the most part—feel numb while they overeat. Until their physical discomfort peaks, at which point they suddenly do feel all kinds of feelings. Guilt. Shame. Remorse. Regret. And so on.

Thus despite the commonly dispensed advice that people overeat to stuff their feelings, in truth, they do it because they aren't feeling—because they're numb. Know it was this knowledge which guided me as I searched for a connection. And sure enough, the connection had been there all along.

The connection? This woman's mother had died the week before that Thanksgiving meal. And until that meal, this woman had been in shock—she'd been numb. Indeed, she remembered feeling numb as the meal began.

What brought on her bulimia? At some point during that meal, she'd eaten enough to suddenly bring her out of shock. And in this moment, she felt—for the first time—her feelings about her mother being dead. At which point, she ran to the bathroom—threw up—and the pattern was set.

Can you now see why she'd been relapsing, year after year, at the same time of year? Obviously, it couldn't have been a fear of getting fat. The fear of getting fat knows no seasons. Can you also see the Cartesian Process at work here? First, I set aside any preconceived notions and became curious. Next, I focused entirely on gathering data. Then I hypnotically regressed this woman to the original scene, so I could experimentally verify this data. Then I used my knowledge of human nature to constellated my findings.

Admittedly, there's much about what I did here which I haven't mentioned. Know we'll return to this story in a future chapter.

On What Actually Makes a Food Good For Us
Speaking of food and throwing up, have you ever wondered what makes people claim entire food groups are bad for us? To be honest, this question bothered me for a long, long time. Then at some point, someone sent me a pre-publication copy of a new diet book. Unlike most diet books, I read this one, cover to cover. Why? Because it intrigued me. It claimed the key to weight loss was to eat foods which were high in water soluble fiber.

Admittedly, at first, I thought, oh well. Another book which divides foods into good and bad. But ten pages in, I read something which forever changed my ideas about food. The authors wrote that eating more fiber is *not* what causes the weight loss. Rather, this fiber is only an indicator that eating these foods can help you lose weight.

Do you get the difference? At first, I didn't. And this is not surprising. While this idea is deceptively simple, our brains simply have no container in which to store it. In truth, patterns like this one are the very opposite of what we're normally told. We're normally given reasons why a diet will cause us to lose weight. But an indicator is not a reason. An indicator is one star in a constellation.

In truth, even after this idea caught my eye, it still took me awhile to realize why. At times, I've even wondered if the authors themselves realized the importance of what they said. I say this as in the final version, they moved this paragraph to the end of the book. Yet to me, this idea should have been on page one.

The idea again? That foods containing nutrients aren't directly good or bad for us. Rather, nutrients and such are indicators—not causes—of a food's benefit to us. This fits with something author Michael Pollen discussed in one of his books on food. He questioned a belief called "nutritionism"—the idea that it's the scientifically identified nutrients in foods that keep us healthy. And if you think about it, we've all been taught that the worth of a food is the sum of its individual nutrients, vitamins, and other components.

Nutritionism is why most packages in supermarkets have Nutrition Facts labels. And if you think back to the last time you visited a supermarket, I bet you read a quite a few of them. So let me ask you, how many times did something on one of those labels influence what you bought? Ironically, those labels list many food facts. They may also lead us to believe—based on those facts—that eating those foods will make us healthy. But does eating nutrients lead to health? Or are nutrients merely indicators—and not causes—of healthy food?

Food for thought, eh?

One Final Thought on Calories

Toward the end of writing this chapter, I heard an interesting talk radio show. And while it focused mainly on weight loss, my ears perked up during one small part. The person being interviewed spoke about the falsehoods inherent in counting calories. And one "falsehood" I'd not heard before involved an idea we discussed in this chapter—the idea that the rate of absorption of whatever we eat changes a food's affect on our weight.

Specifically, this person mentioned the idea that not all calories are equal—that some absorb differently than others. This made me ask, should we be counting calories which absorb more quickly, differently than calories which absorb slowly? For instance, perhaps the calories in the foods I call, "pre-digested" should have a greater "weight gain index?" Whereas the calories in natural foods high in water soluble fiber should have a lower "weight gain index?" Also, should we overlook calories which we do not absorb (e.g. insoluble fiber) as far as weight gain and loss?

Admittedly, I'm still processing these questions. But I'm intrigued.

Chapter 12 - Deafness

On "Mental Sound" vs "Inner Voice" and "Inner Speech"

During this chapter's data gathering step, I found many references to the terms, "inner voice," and "inner speech." Yet despite these references, I chose to use the term, "mental sound." Know I chose to do this because these *inner voice* and *inner speech* references either refer to hearing language or to hearing intuition. Whereas I've used the term *mental hearing* to refer to all sounds heard in the mind.

Why mention this difference? Because like most references in current scientific research, the first two references fail to clearly define terms. For instance, I don't understand why they choose to make a distinction between the mind hearing language and the mind hearing all other sounds. So while I do see the legitimacy in focusing on the aspects of language that involve sound, I cannot see why different variations on this sense are treated as if they are different senses.

My point? These kinds of unnecessary distinctions are at the heart of human prejudice. Deaf people are different—not broken. Sound is sound. And yes, there are obvious differences, including the downsides to being deaf. But why focus only on these downsides when there is much beauty in what Deaf people hear, a beauty which hearing people will never understand.

Are you beginning to see what makes creating measurably-universal definitions so important?

Longitudinal Waves (e.g. sound) vs. Transverse Waves (e.g. water)

In chapter 12, I mentioned how sound traveling in air sort of resembles what happens when you slap the water on the surface of a pool. What I failed to mention is that there are two kinds of waves in play here. The first kind are called longitudinal waves—and the particles in these waves move parallel to the motion of the wave. And the second kind are called transverse waves—and the particles in these waves move at a ninety degree angle to the motion of the wave.

Finally, in case you're having a hard time picturing these ideas, here's another of my over-simplified drawings.

Longitudinal Waves (e.g. sound waves)
vs
Transverse Waves (e.g. water waves)

Longitudinal Waves (the medium moves parallel to waves)
When an impact causes air to move, our bodies convert this movement into sound. Here, sound waves (waves of air) move towards and away from us—in alternately expanding and contracting spheres. In effect, regardless of the angle with respect to the point impact, sound waves expand out from and then compress back parallel to their point of origin.

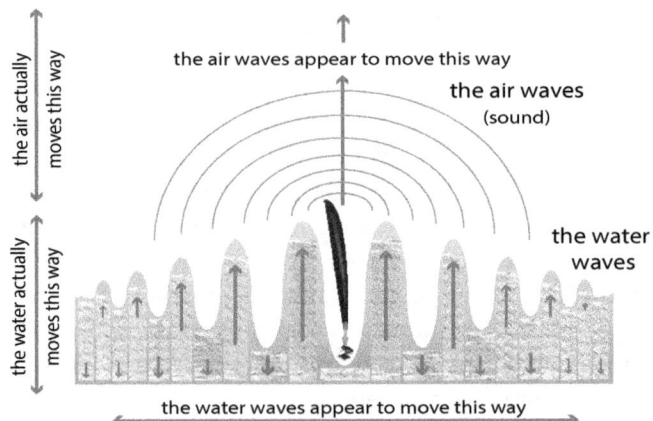

Transverse Waves (the medium moves at right angles to waves)
When a rock splashes into water, it appears that the water waves move horizontally—out and away from the splash. But this is only an illusion. In truth, columns of water "waves" alternately move up and down. To see this for yourself, throw something which floats into a pond—a ball for instance. Now throw a big rock into the water near this ball. The ball will move up and down, but not away from where it started.

On Hearing as Conversions

During the writing of this chapter, I had a conversation with one of my formal students. She happens to have a bachelors in the science of hearing. She is also a teacher of Yoga teachers. When I described the physical to electrical conversions at the heart of all five senses, she immediately told me a similar thing existed in the ancient Sanskrit writings she had read on mandalas as mental sound—and Jedda as the seeds of mental sound.

I mention this as I love it when modern science and ancient teachings parallel each other, just voiced in different forms. If truth itself cannot be ad hoc, then universal patterns must thread through all of time and space.

On Deaf vs deaf (big "D" vs little "d")

In hindsight, I realize I may have inadvertently confused some people with the way I've used the word "deaf." To wit, sometimes, I capitalized this word—sometimes I did not. Small "d" refers to all people who medical professionals designate as being deaf. Large "D" refers to deaf people who identify with the Deaf Community.

Human Sounds Convey Emotions Faster Than Words

Not long after finishing this chapter, I came across a study which claims we pay more attention—and respond measurably quicker—to the emotion in spoken non-word sounds than to the emotion in spoken words.

"Vocalizations appear to have the advantage of conveying meaning in a more immediate way than speech," says Pell. "Our findings are consistent with studies of non-human primates which suggest that vocalizations that are specific to a species are treated preferentially by the neural system over other sounds."

To be honest, the implications here are extremely complex. Certainly, they encompass far more than the ideas contained in this brief paragraph. They also constellate to much of what we've explored in this chapter. Specifically, this study offers evidence which supports the idea that we *physiologically process* sound and meaning in separate areas of the brain, as well as the idea that we also *experientially process* sound and meaning as two separate things. In addition, it implies that we experience sound and meaning sequentially (sound, then meaning).

Know I saw these things even after only one quick read. Sadly, the authors felt the need to posit the usual psychological explanations for their results (e.g. such and such exists because of evolution). I only wish more scientists would see the good in letting their evidence stand on its own. This said, their results support a lot of what we looked at. Thus I highly recommend reading this paper for yourself. (McGill Univ. 2016).

On Musical Ear Syndrome (MES)

When I was researching "mental sound" as something separate from physical sound, I came across MES—a rather odd condition. People who have it suffer from some form of hearing loss. But they at times hear auditory hallucinations—sounds when there is no physical source. Variations on this theme include hearing music when no music is present. In my childhood, my mother experienced this. She once yelled down to me while I was in the basement and told me to lower the music. Being an obnoxious teen at the time, I delighted in telling her there was no music. And in truth, the basement had been silent at the time.

Not surprisingly, during the writing of this chapter, I recalled this event. At the time, the Cartesian Process had opened my mind. Then, in through this opening slipped some pretty strong feelings of remorse, as I recognized how heartless I had been to her deafness—or lack thereof.

In the end, of course, this raised yet more questions in me about deafness and hearing. For instance, is hearing *silence*, "hearing?"

What About Misophonia?

Finally, I want to mention another rare condition—Misophonia; literally, "the hatred of sound." Like many medical conditions, most of us exhibit at least a few symptoms of this one—some sounds annoy us. But when sufferers of the full-blown condition hear sounds they hate, they get mad and or experience fight / flight symptoms (sweating, muscle tension, and quickened heartbeat). What kinds of sounds? Everything from swallowing, gulping, burping, lip-smacking, slurping, throat-clearing, nail-clipping, chewing, tooth-brushing, breathing, sniffing, talking, and sneezing to yawning, walking, drinking, gum-chewing or popping, laughing, snoring, clicking dentures, typing, coughing, humming, whistling, singing, certain consonants, or repetitive sounds.

Why mention this condition? Because it offers yet more evidence for the idea that sound and meaning are separate parts of hearing. In this case, I wonder if the physical sounds even get heard as sounds. Or can it be, these folks experience them only as annoying *noises*?

Chapter 13 - Cancer

On the Fear of Cancer

Of the four chapters in this section, the one on cancer was the toughest to write. In part, this was because—ironically—during the writing, tests led me to believe I might have cancer. Eventually, this was ruled out—the anomalies were being caused by other, far more benign things. But for a time, it left me with a strong desire to bail on the whole book.

Looking back, I realize my fears pushed me to want to avoid thinking about it. Realizing this led me to a ask few questions I had not posed during the chapter. For one thing, can hopelessness be the most devastating symptom of cancer? For another, is hopelessness the greatest factor in most recovery? And for another, is hope the most efficacious form of mental, cancer therapy?

On DNA, Genes, Baseline Pairs, and Logical Geometry

Something I failed to mention in the chapter has to do with DNA. So while I did refer the role DNA plays in cancer, I failed to say anything about its structure. DNA is literally made up of four baseline pairs. Thus DNA is a literal example of how logical geometry exists in our bodies. More important, it's actually a good example of how the sixth geometry unfolds—the mathematical sequence wherein One becomes Two (the first pair of complementary opposites), and Two becomes Four (each pair of complementary opposites then pairs off with a second pair of complementary opposites.) Here's what I mean.

Most people know our DNA contains an overarching set of biological instructions. Everything in us is based on this single set of instructions. This makes this DNA molecule a single point. This point is the One.

Most people also know that each DNA molecule is made up of two intertwined DNA strands (the double helix). Here, each of our parents contributes one strand. Taken together, these two stands are the One becomes Two. This is the first instance of the third geometry.

At this point, things really get complicated, as each of these DNA strands is composed of two pairs of pairs [breathe now]. This is where the Two becomes Four (the first instance of the fifth geometry). Here the first pair is the *pyrimidine* nucleobase pair: cytosine (C) and guanine (G). And the second pair is the *purine* nucleobase pair: adenine (A) and thymine (T). And each nucleobase in one strand bonds to a partner in the other strand. Here, A's can only bond with T's—and G's can only bond with C's.

Now as if this is not enough logical geometry, there is yet another complementary pair. This pair describes the two kinds of sequences which occur in DNA coding. Formally, this pair is known as the sense / antisense sequence pair. And these pairs exist both within single strands of DNA and between the paired strands. Yet another instance of the third geometry.

Has your mind shut down yet? If so, I'm not surprised. The thing is, we're not done yet. It turns out, DNA contains a measurable tipping point—DNA with high GC content is more stable than DNA with low GC content. In addition, yet another tipping point exists in how tightly or loosely the two DNA strands are entwined. Strands that completely circle the axis of the double helix in 10.4 base pairs are sitting on the theoretical tipping point. But in the real world, circles occur in tighter or looser circles. Thus they either tip towards being held more tightly or loosely together. The tight direction is called *positive supercoiling*, and the loose direction is called *negative supercoiling*—and with *negative supercoiling* the bases come apart more easily, and visa versa.

Okay. Enough already. Even my head is swimming. Obviously, what would help would be to constellate all these relationships in a map. To be honest, I'm so curious right now, I'm tempted to do just that. Good thing, I'm no chemist. Still, what I can see just from creating this limited version of the Cartesian Process is this:

[1] The diagram would need to be titled DNA.

[2] One of the map's two questions might have something to do with which parent a strand came from.

[3] The four nucleobases (A, T, G, C) would be either answers, classes of discovery, or natural states. Here, nucleobase A would need to be placed opposite T, and Nucleobase G would need to be placed opposite to C.

[4] The sense / antisense pair would need to be another third geometry. Could this be the second question?

[5] Would positive and negative supercoiling be the vertical axis?

My point for all this? Even this quick and dirty review of DNA points to how common it is to find nested iterations of naturally occurring, logical geometry.

Misc

On Focusing the Search, and On Discovering Baseline Pairs

Speaking of logical geometry, one idea I can't mention enough is the one I just mentioned yet again—the idea of *baseline pairs*. These amazing sets of natural opposites have the power to open up whole new lines of questioning. Moreover, since developing new lines of questioning is one of constellated science's main goals, obviously, finding these pairs is important. No surprise then that I've focused so much in this section on this one idea.

This said, I have to admit, it can sometimes be hard to find a baseline pair. For instance, with insulin, I haven't a clue where to find one yet. Ironically, with the insulin-centered condition—diabetes—the pair had already existed—Type 1 and Type 2. And thankfully, throughout this section, we've been able to identify baseline pairs for key concepts.

We've done this for sleep states: Type 1 (nonREM / body free & mind frozen) vs Type 2 (REM / mind free & body frozen). We've done this for extreme weight change states: Type 1 (anorexia) vs Type 2 (bulimia). We've done this for deafness: Type 1 (no structural damage / sudden onset hearing loss) vs Type 2 (structural damage / slowly increasing hearing loss). And we've done this for cancer: Type 1 (rapidly dividing, undifferentiated cells) vs Type 2 (slowly dividing, losing their differentiation cells).

This raises the question as to how many medical conditions might better be described were we to categorize them as Type 1 / Type 2 pairs. Here, Type 1 would roughly refer to cases which involve damaged instructions leading to unnatural cellular creation, and Type 2 would roughly refer to cases which involve damage to normal cells.

So for example, would Rheumatoid arthritis be Type 1 and would Osteo be Type 2? This way of typing them seems to be similar to how we've typed cancer and deafness. Then again, I have to admit—I sorely lack a scientific way to define the two Types. And yes, I know that together, they are far more than mere opposites. But what do all the Type 1's have in common that Type 2's don't contain? For instance, is the speed of onset a factor. Do all Type 1's have a rapid onset, so you can't see them coming vs Type 2's have a slow onset, so you can see them coming?

Hopefully, the science for this will emerge over time.

On Babies as Constellated Scientists

Finally, I want to mention one last thing—the idea that babies practice most if not all of what we just talked about from birth to about age two. In a way then, all babies are little constellated scientists—which in part explains why they constantly move about, renewing their curiosity. And while they lack the skills to make lists, it seems, they don't need them—the world out there is their never-ending list.

The big thing of course is to realize how babies employ logical geometry in everything they do. Pattern recognition is their primary tool—mapping their world is their primary goal. Amazingly, while babies do occasionally get frustrated, they never quit, blame, or pontificate. Moreover, while they sometimes struggle to share physical possessions, they constantly seem open to sharing their discoveries.

Imagine a world in which scientists functioned like this?

Resources for Chapters 10 - 13

On the Science of Sleep

As I've acknowledged throughout this section, gathering data is an essential element in all scientific endeavors. The following books and online sources supplied some of my data in the chapter on sleep.

Kryger, Meir H., T. Roth, W.C. Dement. (2005). *Principles and Practice of Sleep Medicine.* Phila, PA: Saunders Elsevier. (Should you opt to read this book, know you'll get extra points for simply lifting this tome. It contains ginormous amounts of data to sort through. Talk about TMI. Unfortunately, it also contains interesting, odd, and typical omissions regarding patterns of connections within and between this data. To wit, most of this data is grouped in separate categories—and no connections between data groups are mentioned.)

Kryger, Meir H. (2010). *Atlas of Clinical Sleep Medicine.* Phila, PA: Saunders Elsevier.

Lee-Chiong Jr., Teofilo. (2008). *Sleep Medicine, Essentials and Review.* New York: Oxford University Press. (An exceptionally well organized book.)

Billard, Michael (editor/translator - contains 73 authors, originally in French). (2003) *Sleep, Physiology, Investigations, and Medicine.* New York: Kluwer Academic / Plenum Publishers.

Naiman, Rubin R. Ph.D. (2009). *Healing Night, The Science and Spirit of Sleeping, Dreaming, and Awakening.* Minneapolis: Syren Book Company. (Personal stories and caring advice).

McKenna, Paul. (2009). *I Can Make You Sleep.* New York: Sterling Publishing. (Self-help with a few surprisingly useful connections. I'm generally skeptical of such books. But I'm glad I bought this one.)

Somer, Elizabeth, M.A., R.D. (1999). *Food & Mood.* Henry Holt & Company, LLC. (Pg. 143) (Connects Serotonin / Melatonin levels to seasons).

I read in several places that the positioning they use in hypnograms is an arbitrary choice. Here's one. (http://www.sleepdex.org/hypnograms.htm, retr. 7/11/15)

Then there's the stuff on serotonin / melatonin, and CO_2, and such.

National Sleep Foundation. (2014, Jan.). Melatonin and sleep. http://sleepfoundation.org/sleep-topics/melatonin-and-sleep, retr. 6/12/15)

[Usha Satish,[1] Mark J. Mendell,[2] Krishnamurthy Shekhar,[1] Toshifumi Hotchi,[2] Douglas Sullivan,[2] Siegfried Streufert,[1] and William J. Fisk[2]. 2012. *Is CO_2 an Indoor Pollutant? Direct Effects of Low-to-Moderate CO_2 Concentrations on Human Decision-Making Performance.* Environmental Health Perspectives, vol. 120, number 12, December 2012]

Finally, the "contempt prior to investigation" quote I used came from: http://www.thefix.com/content/contempt-prior-to-investigation-AA-Herbert-Spencer8042.

On the Science of Weight Loss

Seale, Stuart A. MD, Teresa Sherard, Diana Fleming. (2010). *The Full Plate Diet, Slim Down, Look Great, Be Healthy*. Austin, TX: Bard Press. (Some may be surprised by the fact that in a chapter on weight loss, I've mentioned only one diet book. Know I've included it not because this diet guarantees success but rather, because with a single idea, it changed my views on food. The idea? That fiber (and by extension, nutrients) are indicators—rather than causes—of healthy food.

It also does not break food into good and bad groups. Nor does it make false or sensational claims. Finally, it has much good information delivered in a common sense style. And in case you missed it, I highly recommend this book.)

Pollen, Michael. (2006) *The Omnivore's Dilemma: A Natural History of Four Meals*. New York: Penguin Press.

Pollen, Michael. (2008). *In Defense of Food: An Eater's Manifesto*. New York: Penguin Press.

Pollen, Michael. (2009). *Food Rules: An Eater's Manual*. New York: Penguin Press.

(What I like best about Pollan is that he offers well-researched data, personal experiments, personal thoughts, and tons of questions. At the same time, he never offers answers. Most important of all, his chapters almost always end with new lines of questioning. Thus I usually leave his books with a strong desire to know and seek more.

He also repeatedly questions beliefs and their effect on our health and our weight. Moreover, if the authors of *The Full Plate Diet* are right, then the main value in nutrients may be that they point to beneficial foods. In other words, perhaps it's not the nutrients or vitamins themselves that make us healthy? Perhaps it's eating foods that by nature contain them?

So is *nutritionism* a flawed belief? It's Pollen who got me to ask this question. Is this why taking mega vitamins doesn't always lead to the improvements we expect? It's been proven again and again that if we seriously lack a vitamin or mineral, that taking vitamins makes up for this lack. But what if we don't have this lack? What then? Does taking bare nutrients benefit us all? Or is it as some researchers tell us; that unless we're deficient, they don't do us much good? Questions. Questions. Questions.)

As for the shortcomings of diets, here's one link:

Harriet Brown. (3/24/15). *The Weight of the Evidence, It's time to stop telling fat people to become thin.* Slate http://www.slate.com/articles/health_and_science/medical_examiner/2015/03/diets_do_not_work_the_thin_evidence_that_losing_weight_makes_you_healthier.3.html) (According to them, "It's hard to think of any other disease—if you want to call it that—where treatment rarely works and most people are blamed for not 'recovering'." I so agree.)

There's also the idea that you must keep reinventing your exercise routines. According to this article, your body adapts to your fitness routine—so over time, you won't burn as many calories as you did when you started your working out. They suggest you alternate your workouts and find fresh movements and fresh exercise formats if you want to continue to burn a specific number of calories on a regular basis. Excellent idea.

Jessica Smith. (2016). *5 Reasons Losing Weight Fast is Easier than Losing Weight for Good And 5 simple tips for long-term success.* http://www.shape.com/weight-loss/weight-maintenance/5-reasons-losing-weight-fast-easier-losing-weight-good/page/2 .

Then there is Gary Taub's book. Taub, Gary. (2007). *Good Calories, Bad Calories, Challenging the Conventional Wisdom on Diet, Weight Control, and Disease.* New York. Alfred A. Knopf. I cannot say enough about this well written, well organized researcher cum story teller extraordinaire. Sadly, many critics panned him, and I can see why. He constellates science in such a way as to force people to question their long assumed beliefs. And who wants to believe their life-long beliefs are false and wrong.

Finally, I want to credit a man whose diet book I have not mentioned. His name is Jon Gabriel. His book—Gabriel, Jon. (2008). *The Gabriel Method: The Revolutionary DIET-FREE Way to Totally Transform Your Body.* New York. Atria Books / Simon & Schuster. I mention this book as it was Jon's website home page that got me to question whether weight loss and gain were rooted in tipping points (e.g. my ideas; Naturally Thin vs Naturally Fat). I also respect him as a genuinely caring, non blaming, common sense human being. And while some of his suggestions are based in will-powered rather than natural change, still, his work is definitely worth reading.

On the Science of Deafness

Denworth, Lydia. (2014). *I Can Hear You Whisper, An Intimate Journey Through the Science of Sound and Language.* New York: Dutton (Penquin)

(I cannot recommend this book enough as well. It is the rare combination of clear science, personal storytelling, and the drive a mother can feel while trying to discover the science necessary to help her child.)

Bouton, Katherine. (2014). *Shouting Won't Help, Why I and 50 Million Other Americans Can't Hear You*. New York: Picadore. (Also a good read. It has similar qualities to Denworth's book, but told from the perspective of an adult facing her own deafness.)

Holcomb, Thomas K. (2013). Introduction to American Deaf Culture. New York: Oxford. (Another must read.)

Meadow-Orlans, K, P. E. Spencer, L. S. Koester. (2004). *The World of Deaf Infants, A Longitudinal Study*. New York: Oxford University Press (An amazing compendium of data regarding the lives of deaf infants. Especially interesting were the comparisons between infants being raised by deaf parents, infants raised by hearing parents, and infants raised by one deaf and one hearing parent.)

Kral, Andrej, A. N. Popper, R. Fay (editors). (2013). *Deafness*. New York: Springer (A seriously technical and articulate tome. For me, too much information with no constellations—but great for gathering data.)

McGill University. (2016, January 18). Human sounds convey emotions clearer and faster than words. ScienceDaily. Retrieved January 24, 2016 from www.sciencedaily.com/releases/2016/01/160118134938.htm (I mentioned this article in the text. Well worth reading and probably contains far more constellations than I found in my brief read.)

http://www.hearingreview.com/2015/05/researchers-discover-brain-reorganizes-hearing-loss/#sthash.4sNSYUHV.dpuf (one source for "cross-modal" cortical reorganization.)

Helena Örnkloo, Claes von Hofsten (2007). Fitting objects into holes: On the development of spatial cognition skills. Developmental Psychology, 43 (2), 404-416 DOI: 10.1037/0012-1649.43.2.404

"The hearing areas of the brain shrink in age-related hearing loss," said Sharma. "Centers of the brain that are typically used for higher-level decision-making are then activated in just hearing sounds. These compensatory changes increase the overall load on the brains of aging adults. Compensatory brain reorganization secondary to hearing loss may also be a factor in explaining recent reports in the literature that show age-related hearing loss is significantly correlated with dementia." - See more at: http://www.hearingreview.com/2015/05/researchers-discover-brain-reorganizes-hearing-loss/#sthash.4sNSYUHV.dpuf

On the Science of Cancer Research

One thing to know about cancer research is that in many ways, cancer is like alcoholism. By this, I mean that to know it, you must research every aspect of human nature. In a book which focuses on describing a new scientific method, this was just not feasible. Indeed, I feel quite overwhelmed by the thought of having to list my sources—I literally referenced hundreds. This said, here are but a few of those I used.

Taub, Gary. (2007). *Good Calories, Bad Calories, Challenging the Conventional Wisdom on Diet, Weight Control, and Disease.* New York. Alfred A. Knopf. (Why list this book twice? Because Taub's book is such a good example of constellated science. From the title, you'd think the book would limit itself to food. But Taub constellates many ideas—and much research data—to connections between diet and food and how it connects and doesn't connect to cancer.)

The Niche; Knoepfler lab stem cell blog. (2012, April) *Who really discovered stem cells? The history you need to know.* https://www.ipscell.com/2012/04/who-really-discovered-stem-cells-the-history-you-need-to-know/

Hebrew University of Jerusalem. (2008, December 22). *Mechanism That Triggers Differentiation Of Embryo Cells Discovered.* ScienceDaily. Retrieved January 31, 2016 from www.sciencedaily.com/releases/2008/12/081221220328.htm

Library of Congress; Everyday Mysteries. (2010, August 23). *What are stem cells?* Retrieved January 31, 2016 from http://www.loc.gov/rr/scitech/mysteries/stemcells.html

Stadtfeld, Matthias and Konrad Hochedlinger (2010). *Induced pluripotency: history, mechanisms, and applications.* http://www.genesdev.org/cgi/doi/10.1101/gad.1963910. Cold Spring Harbor Laboratory Press.

Eifrig Jr., Dr. David, MD. (2014). *The Living Cure, The Promise of Cancer Immunotherapy.* Baltimore, MD: Stansberry Research (Not much of a book really, more like a draft of a college paper. I include it because it exists in an area where there are no books. It does include a few good general viewpoints, and a decent bibliography. But you might want to pass on this one.)

Misc

One final acknowledgement. I cannot begin to thank the two authors whose videos on scientific method I studied during the course of writing this book. Both men are published by The Teaching Company—professors Steven L. Goldman (Science Wars) and Jeffery L. Kasser. (Philosophy of Science) Together, these two men offer an awesome resource for the nature of scientific method. Moreover, both men have challenged me to discover more.

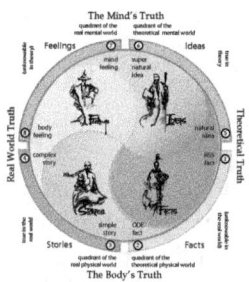

Section 3 - Introduction

The Mind
(what can we learn about...?

Oops, Did I Do It Again?

By now, I'm sure it's obvious, I've overreached once again. We're well past the 100,000 word mark. Ordinary books average 50,000 to 70,000 words. Still I think we've accomplished a lot and hopefully, some of it will make a difference. But before we go, I'd like to tell you what I'd planned to put into this book—and what I hope to include in the next.

Originally, this book was to have four sections—I've managed to make it through two. In section one, I spent three chapters formally introducing you to Logical Geometry—the mathematics behind the new method. Then in section two, I introduced you to constellated science's Cartesian Process by exploring four topics—Sleep, Weight Loss, Deafness, and Cancer.

What I didn't get to were the other two sections. In section three, I planned to introduce you to the fourth stage of constellated science—the constellating stage. And in section four, I wanted to offer an example of how constellated science might raise the bar for a whole profession.

As for specifics, I'd titled section three, **The Mind (what can we learn about...?)**. Here, I planned to explore four more topics.

- In **Chapter 14**, I planned to explore The Science of IQ, Genius, & Creativity. Can we alter IQ? Can we teach *genius*? Can children permanently be infused with creative ability?
- In **Chapter 15**, I wanted to explore The Science of Learning & Natural Change. Can we teach children without forcing them to parrot answers? Can we scientifically test for learning? Do we really forget what we've learned?
- In **Chapter 16**, I was going to explore The Science of Emotion & States of Mind. Can we scientifically define emotions? Is algebra the key to understanding the mind? Can we quantify and or separate individual thoughts and feelings?
- In **Chapter 17**, I planned to explore The Science of Love & Relationships. Are there knowable, natural patterns which underlie human relationships? Could knowing them make us better parents? Could they teach us to love?

~

Looking at these titles, I admit, any one of them could easily become a whole book. And this doesn't include section four, which I'd titled, **A Whole Profession (what can we learn about . . . ?)**. The profession? Talk therapy—and the healing professions in general. Not surprising, really. I am after all a talk therapist.

Know I've wanted to scientifically explore talk therapy for almost a decade now. Why? Because this profession may hold the key to one of science's more important, unsolved mysteries—how to define "wounds." Amazingly, science has never actually defined this word. And yes, they often say it's the symptoms. But getting rid of symptoms doesn't guarantee they won't come back. And since you can literally watch a film of a painful event and the symptoms will persist, a wound cannot be a lack of knowledge as to the details of—and reasons for—these painful events.

What would we gain if we could scientifically define *wounds*? Just consider what we've explored in this book. Cancer, deafness, weight loss, and sleep problems. All would be affected. Then there are our difficulties with learning, with emotions, with relationships, and with love. In truth, there's nothing this definition would not help us to improve.

Know that prior to writing this book, I would have felt pressured to speculate on what we might find. Today, I feel more able to comfortably remain in a state of unsatisfied curiosity. No doubt that repeatedly advising you, the reader, to focus on questions rather than on answers, has affected me too. Including my final question. Is it possible the new method is the path to the next great, scientific leap? I so want to know.

www.ingramcontent.com/pod-product-compliance
Lightning Source LLC
Chambersburg PA
CBHW050618300426
44112CB00012B/1554